Mammals
of South America

Mammals
of South America

Rexford D. Lord

THE
JOHNS HOPKINS
UNIVERSITY PRESS
BALTIMORE

© 2007 The Johns Hopkins University Press
All rights reserved. Published 2007
Printed in China
on acid-free paper

9 8 7 6 5 4 3 2 1

The Johns Hopkins University Press
2715 North Charles Street
Baltimore, Maryland 21218-4363
www.press.jhu.edu

Library of Congress Cataloging-in-Publication Data
Lord, Rexford D.
   Mammals of South America / Rexford D. Lord
     p. cm.
   Includes bibliographical references and index.
ISBN 0-8018-8494-2 (hardcover : alk. paper)
1. Mammals—South America. I. Title.
QL725.L67 2006
599.098—dc22   2006007083

A catalog record for this book is available
from the British Library.

CONTENTS

PREFACE

South America can boast of possibly having the greatest number of species of birds (about 30%) and mammals (about 20%) of all the continents. It is a privileged continent. It spans both sides of the equator and contains both the world's longest mountain chain, the Andes, and largest (in volume) river, the Amazon. The Amazon jungle surpasses all other tropical forests in size. Species richness in South America is remarkable, deriving from the diversity of its life zones and habitats, yet this amazing continent cannot boast the magnificent large mammals found in Africa.

The largest mammal of South America, the tapir, is small by comparison with the rhinoceros or elephant and its largest feline predator, the jaguar, is smaller than either the lion or tiger. What South American mammals lack in size, they make up for in strange and wonderful forms. Although Australia is the land of marsupials, South America has its fair share of opossums. Armadillos are found only in the Americas with a total of 21 species; all are found in South America, with one extending its range to North America. (Although the pangolins of Asia and Africa resemble armadillos, they are not related.) The anteaters and sloths are distinctive mammals that are unparalleled anywhere.

Extensive grasslands border the Amazon Rain Forest, with the llanos to the north and the pampas to the south. Both plant and animal species vary greatly between these grasslands, but many species of mammals and birds are found in both. In contrast to North America, South America tapers to a point as it approaches the South Pole. Consequently, its climate is more maritime and does not produce the extreme cold of the winters of North America at equivalent latitudes. The southern half of the "Southern Cone," a land called Patagonia, is a region beyond compare. There, the wind blows day and night, stopping briefly only at dawn and dusk. The power of this constant wind is so great that it shows on aerial photographs as east–west streaks in the vegetation pattern.

Because of the differences in climate and topography, a great contrast exists between the mammals of North America and South America. This contrast is greatest when the division is made between temperate North America and tropical South and Central America, but including, of course, temperate South America. In other words, by using biogeographical rather than the usual political divisions, a natural dichotomy becomes evident.

There are three distinctive South American mammal groups: the marsupials, the xenarthrans, and the hystricognath rodents. Only one species of each group has invaded North America, the Virginia opossum (*Didelphis virginianus*), a marsupial; the nine-banded armadillo (*Dasypus novemcinctus*), a xenarthran; and the northern porcupine (*Erethizon dorsatum*), a hystricognath rodent. A few other South American groups can also be thus considered. For example, the North American raccoon (*Procyon lotor*) is alone in temperate North America but it has several relatives in the tropics. There are two species of manatee in South America, but only one gets up to the United States, the West Indian manatee (*Trichechus manatus*). Likewise, two species of dolphin live in the fresh waters of South America, but none are found in the fresh waters of North America. There are, of course, the American camels, the llama, alpaca, vicuña, and guanaco, whose ancestors came from North America in the geological past but have no native representatives there today.

Apparently, only a few North American mammal groups have been somewhat inhibited in their invasion of South America. Indeed, only one bear, the spectacled bear (*Tremarctos ornatus*), is found in South America. Likewise the shrews have only one genus (*Cryptotis*) in South America, and it is confined to the Andes Mountains.

A spectacular example of a species that has spanned both North and South America is the puma, or mountain lion *(Puma concolor)*. This magnificent large cat is found all the way from Canada to Patagonia, living in mountains and lowlands, in the humid tropics as well as deserts.

It is relatively easy to see how tropical forms most abundant in South America have occupied the tropics of Central America and Mexico. Tropical bats such as the common vampire bat (*Desmodus rotundus*) and many other species of the leaf-nosed Phyllostomidae fall in this category. Likewise many marsupials, xenarthrans, and hystricognath rodents have spread northward.

North Americans are sometimes surprised to find some northern species also distributed in South America. Among such species are the cottontail rabbit (*Sylvilagus floridanus*), the gray fox (*Urocyon cinereoargen-*

teus), and the white-tailed deer (*Odocoileus virginianus*). All three are abundant in northern South America.

If the mammalian species from North and South America were to compete, the South would win. There are more than a thousand species of mammals in South America. The number of species in North America does not approach that. In North America (north of Mexico) almost all bats (40 some species) are insectivorous. In South America bats of many families and species (more than 200 species) consume a much wider variety of items: insects, of course, but also fruit, pollen, nectar, birds, frogs, and blood.

An interesting example of convergent evolution has arisen between North and South America. Found in North America is a group of burrowing rodents possibly related to squirrels called pocket gophers. Pocket gophers comprise 35 species, belonging to five genera. They spend most of their lives burrowing beneath the ground foraging on roots and tubers. In South America there is a group of completely unrelated rodents called tuco-tucos that also spend their lives beneath the ground, foraging on roots and tubers. The tuco-tucos are so named for their calls, which are audible above ground.

Paleontological evidence indicates that many South American mammal groups came from North America, but so long ago that these groups evolved their specialized forms in South America and no longer resemble their progenitors. Since those early invasions, other invasions have also occurred, most notably of rats and mice. They invaded South America during the late Oligocene or early Eocene, traveling across the Panamanian land bridge and rapidly dispersing and evolving species to fill all available niches. Curiously, the squirrels, having invaded from the north where they abound in cold climates, all live in tropical forests in South America, not in the more temperate climes high in the Andes Mountains. Perhaps, having adapted to the tropics in Central America, those moving into South America lost their ability to adapt to colder life zones.

* * *

This book follows the latest taxonomic terminology and organization established in the third edition of *Mammal Species of the World* (Wilson and Reeder, 2005).

To my wife, Veronica. I am grateful for her understanding of my devotion to this task, a work that began before we were married many years ago. Persons who contributed most significantly and directly were the late Arthur Greenhall and Julio Cerda. Charles Handley identified many bat species that were kept in the Smithsonian collection. Leslie Pantin of-

Guanaco (*Lama glama*)

fered important suggestions and kindly permitted me to photograph the animals in his care. Merlin Tuttle, Ronald Pine, William Davis, and others sent me literature while I was living in Latin America. Saul Gutiérrez assisted in numerous ways. Others who helped in various ways are: Alfred Gardner, Don Wilson, Chris Wozencraft, Nancy Simmons, Michael Carleton, Colin Groves, Michael Mares, Robert Timm, Eileen Mathias, Paul Greenhall, Omar Linares, Abel Fornes, Elio Massoia, Horacio Delpietro, Jorge Crespo, Melvin Castillo, Humberto Cuenca Aguirre, Ernesto Gorgoglione, Farouk Muradali, Leopoldo Lapenta, Joseph Gallen, Albino Belloto, Mirta Roses, Fernando Dora, Russell Mittermeier, Stam Zervanos, June Brown, Perry Habecker, and Karl Koopman.

Most of the photographs were taken by me. Others who contributed photographs were Never Bonino, Joseph Gallen, Gloria Jafet, Michael Mares, and Horacio Delpietro. Photographs were also contributed by the Mammal Image Library of the American Society of Mammalogists. Those who contributed photographs through the Mammal Image Library were A. H. Shoemaker, M. A. Rosenthal, T. Carter, R. P. Fontaine, H. Tyndale-Biscoe, J. F. Eisenberg, Justine Anderson, P. Myers, F. A. Cervantes-Reza,

G. E. Svendsen, C. M. Drabek, J. Rood, L. Pinder, P. L. Meserve, G. Hockaby, and J. F. Gray. The following persons contributed in my constant quest for photographs of mammals or in other ways: Horacio Delpietro, Vinicio Sosa, Luz María Aguilar, José Ochoa, Fidel Yunes, Marisol Aguilera, Edgardo Mondolfi, Omar Linares, Heriberto Merchan, Victor Delgado, Mirna Quero, Freddy Bermejo, Haydy Monsalve, Tulio Aguilera, Leopoldo García, Lila Adrián, Alberto Fernández, Ernesto Fernández, Jesus Rivas, Carol Reigh, María Muñoz, Mercedes Sohares, Juan Gómez Nuñez, Dane Burkhart, and Eduardo Lander. The Directors of COVEGAN, Efrain Barroeta, Alcides Gonzalo, Jesus Pacheco, and Enrique Loreto, owners of Hato El Cedral, Apure, Venezuela, were generous in their support of my efforts.

I have been able to photograph many species in zoos. The following zoos openly offered their facilities to me and I am very thankful to all: Pittsburgh Zoo, Brandywine Zoo, Philadelphia Zoo, Zoológico del Parque del Este (Caracas), Zoológico de Bogotá, Zoológico de São Paulo, Zoológico de la Ciudad de México, Zoológico de Toluca (Mexico), Zoológico de Managua, Zoológico de Lima, Zoológico de La Paz, Zoológico de Quito, Zoológico de Maracay (Venezuela), Zoológico de Valencia (Venezuela), Zoológico de Yaceretá (Paraguay), New Orleans Zoo, Hattiesburg Zoo, and Madison (Wisconsin) Zoo.

PHOTOGRAPHY AND MAMMALS

Salvin's big-eyed bat (*Chiroderma salvini*)

This book, through its pictures, puts forth the following premise. Due to a unique feature in the mentality of humans, we can readily recognize subtle differences and similarities in faces. We can see familial similarities and even call them to the attention of others with remarks such as "he is the spitting image of his mother." This ability can be applied as an additional method for recognizing differences and similarities between mammals.

It is not suggested that photographs be used to differentiate between individual animals, but rather that they help in our personal recognition of certain groups, in particular, genera within the bats. South American bats have remarkably varied faces, but because they are small, relative to what we are accustomed to focusing on, photographic enlargement reveals their differences and similarities.

Good artists can draw recognizable faces of well-known personalities. Good artists can also illustrate the facial features so readily apparent in bats' faces, but to do so, they must work from an enlarged photograph. It is useful to create a faithful rendition of the subtle facial nuances as revealed by the variations of light, shadow, and color of the photograph itself. Nevertheless, artists can do what photography cannot do; they can combine the features of several individuals to provide a better rendition of the species.

BIOGEOGRAPHY

The geography of South America is complicated and is responsible, in part, for the great diversity of mammal species. To understand this complicated system it is helpful to isolate the separate biomes—the major ecological units—beginning with the largest, the Amazon Rain Forest. The Amazon Rain Forest is located primarily within Brazil, but includes adjacent portions of Peru, Ecuador, Colombia, Venezuela, and the Guianas. The large grassland biome south of the Amazon Rain Forest is the Pampas, occupying northern Argentina and Uruguay. To the south of the Pampas lies the Patagonian biome, a dry brushland, primarily in Argentina. To the north of the Amazon Rain Forest lies another grassland region, the Llanos biome, which occupies primarily southern Venezuela and southeastern Colombia. The Andes Mountain biome runs north from Tierra del Fuego through Chile and western Argentina, western Bolivia, Peru, and Ecuador to Colombia and western Venezuela. This mountain biome is composed of complex zonations that depend on altitude and latitude as well as orientation with respect to being on the east or the west slope.

The Amazon Rain Forest, the Pampas, the Llanos, Patagonia, and the Andes Mountains form the principal biomes of South America, but important secondary biomes are found beyond and between the principal biomes. Between the Amazon Rain Forest and the Pampas lies the Gran Chaco biome occupying western Brazil, eastern Bolivia, Paraguay, and northwestern Argentina The Chaco is characterized as a dry region with prominent cactus and thorn bush, yet in lower, poorly drained spots it resembles the Pantanal. The southern part of the Chaco (Argentina) tends to be very dry with the soil of country roads converting to deep talc until the wet season returns. The northern part of the Chaco (Paraguay and Brazil) gradually converts to the conditions of the Pantanal (Brazil), with some scattered swampy areas. A significant portion of the Chaco is located in the state of Mato Grosso. This state is possibly named for a distinctive tree "Palo Borracho" (*Chorisia ventricosa*). It is commonly seen

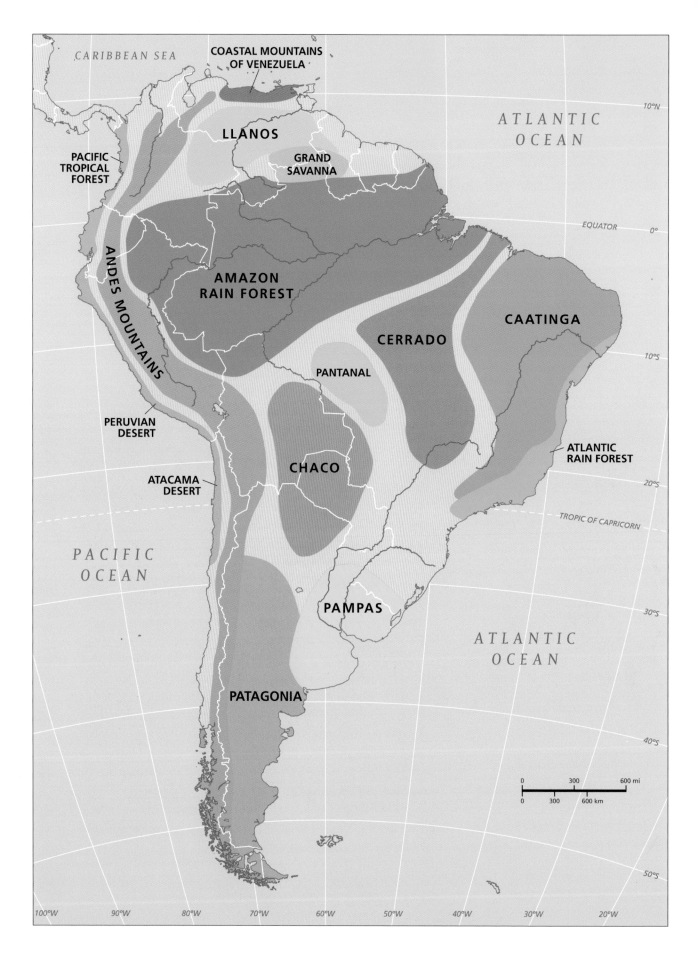

CARIBBEAN SEA

COASTAL MOUNTAINS
OF VENEZUELA

LLANOS

GRAND
SAVANNA

PACIFIC
TROPICAL
FOREST

ATLANTIC
OCEAN

10°N

EQUATOR 0°

ANDES MOUNTAINS

AMAZON
RAIN FOREST

CAATINGA

CERRADO

PANTANAL

10°S

PERUVIAN
DESERT

ATLANTIC
RAIN FOREST

ATACAMA
DESERT

CHACO

20°S

TROPIC OF CAPRICORN

PACIFIC
OCEAN

PAMPAS

30°S

ATLANTIC
OCEAN

PATAGONIA

40°S

0   300   600 mi

0   300   600 km

50°S

100°W   90°W   80°W   70°W   60°W   50°W   40°W   30°W   20°W

throughout the Chaco. It is readily recognized by its distended rotund trunk, yet its leaves and flowers are similar to the related silk cotton tree (*Ceiba pentandra*), so common to the more humid American tropics.

In Brazil, southeast of the Amazon Rain Forest lie two important secondary biomes, the Cerrado and the Caatinga, both completely within Brazil. The Cerrado is a region with marked wet and dry seasons centered on the states of Goias and Tocantins. Its primary vegetation is deciduous trees. The Caatinga is located in the northeastern region of Brazil, centered on the states Pernambuco and Paraíba. It is a semiarid region vegetated with short trees and tall bushes. Between the Cerrado and the Chaco lies the Pantanal, a very swampy region. South of the Caatinga, hugging the coast, is the Atlantic Rain Forest found in the states of São Paulo and Paraná. A transition zone lies between the Pampas to the south, the Chaco to the west, and the Amazon Rain Forest and Cerrado-Caatinga to the north, occupying Paraguay and parts of Brazil, Bolivia, Argentina, and Uruguay.

Another secondary biome, the Grand Savanna, lies to the east and south of the llanos of Venezuela and extends into Guyana. It is a region of impoverished soils with very few trees. The Guianas likewise form a tran-

Llanos

Patagonia

Andes Mountains, Mt. Cotopaxi

sition zone between the Amazon Rain Forest, the Grand Savanna, and the Llanos, and the Coastal Mountains of northern Venezuela constitute yet another secondary biome. A semidesert lies in northwestern Venezuela. The driest desert in the world, the Atacama, lies west of the Andes Mountains on the Pacific coast in northern Chile. The entire coast of Peru likewise is desertlike, but from Panama south along the Pacific coast of western Colombia and including northwestern Ecuador is found a region of humid tropical forest.

Although these biomes are readily recognized by their characteristic plant life, their plant and mammal distributions do not exactly coincide. Nevertheless, because mammals depend on plants, each biome contains a characteristic mammalian fauna. There is a tendency by some to think of the Llanos and the Pampas as similar grasslands, merely separated by the Amazon Rain Forest. The two are quite different, in particular, in soil fertility. The Pampas has regions of soil rich enough to serve international grain companies in the production of hybrid grains for exportation. The Llanos soil is so leached of nutrients, on the other hand, that even attempts to grow better grasses as cattle forage require the addition of fertilizers. The Pampas lies in a temperate climate, whereas the Llanos is in a tropical environment. Both experience a wet season and a dry season annually, but the Llanos varies from being flooded to being parched, and the Pampas exhibits a more moderate variation in humidity.

Mammals
of South America

1. Marsupials

Marsupials are found in South America, Australia, and some islands of Oceania. Many anatomical features differentiate the marsupials in South America from those in Australia, but the most notable is the number of incisor teeth. South American marsupials have more incisor teeth, five upper incisors and four lower incisors, while other marsupials, such as the kangaroos, have even fewer incisor teeth. Another notable difference is the prehensile tail (one that can grasp branches). South American marsupials have prehensile tails, whereas only a few non-American marsupials have such tails.

Bare-tailed woolly opossum (*Caluromys philander*)

The Order Didelphimorpha is a group of marsupials that is found only in the Americas. It comprises 17 genera and 72 species called the "American opossums." The distinctive marsupial pouch is present in the larger species of this group, but it is often absent or rudimentary in the smaller species. The tails of American opossums are usually long, bare (at least toward the tip), and prehensile. The ears are usually large and bare. Most are great climbers, with the exception being the genus *Monodelphis,* members of which can climb but seem to prefer the ground. Even the arboreal species spend at least part of their time on the ground. All the feet of American opossums have five toes, with the inner large toes of the hind feet usually being opposable (like the human thumb) for grasping tree limbs. This "thumb" lacks a claw.

The gestation period is short, a matter of 12 to 24 days. The young are born in an almost embryonic form, with their eyes closed and their hind limbs merely stubs. In contrast to the hind limbs, the forelimbs are well developed at birth, with claws on their digits to aid them as they make their journey to the pouch. The mother's teats, up to 13 (usually an odd number of nipples), are located within the pouch. The young depend on milk for their subsequent growth until weaning. When small, the young are always found within the pouch, attached to the teat. As they grow they often cling to their mother's body. In the pouchless species, the young simply cling to the body, taking nourishment from the mother as needed.

A drawing showing a mother opossum with her tail held over her back, with several young hanging by their own tails from her tail, often appeared in some older books. This was a false depiction. The young either stay in the pouch or cling to the mother's fur or skin with their claws.

South American marsupials are remarkable in their achievements. They are generalists that can live in a great variety of habitats, from tropical forests to deserts and from sea level to high in the mountains. Their skulls reveal that they have a very small brain, which lacks the corpus callosum, the central part of the brain connecting the halves. Yet with this small, primitive brain, they can outwit humans, stealing food and even chickens from their coops during the night. American opossums of the genus *Didelphis* (the most abundant marsupials in South America) are slow moving and appear none too agile. They often feign death if they are attacked and bitten by dogs. They exude a foul-smelling greenish substance from their rectum when they feign death, discouraging further aggression and saving them from death. They lie there with the bad smell, often attracting flies, which walk over the carcass looking for sites to lay their eggs. Then, after perhaps 15 minutes, they get up and walk away, apparently none the worse for the experience.

Brown-eared woolly opossum
(*Caluromys lanatus*)

Bare-tailed woolly opossum
(*Caluromys philander*)

Black-eared opossum
(*Didelphis marsupialis*)

Robinson's mouse opossum
(*Marmosa robinsoni*)

Woolly mouse opossum
(*Micoureus demerarae*)

Northern red-sided opossum
(*Monodelphis brevicaudata*)

Common thylamys (*Thylamys pusilla*)

The order Paucituberculata contains a single family, the "shrew opossums" in the family Caenolestidae. The pouch that typifies the marsupials is absent from this group, although a rudimentary pouch is sometimes observed in young animals.

The shrew opossum family is composed of six species that are restricted to western South America. These are truly primitive marsupials. They are shrewlike in appearance with pointed muzzles, and they live high in the damp Páramo of the Andes Mountains (a wet temperate grassland above the tree line). Uniquely, the tails of shrew opossums are covered with stiff, short hairs. Like the American opossums, they have five toes on all four feet. The first and fifth toes are quite small, however, and the claws on their forefeet are flattened, while the well-developed middle toes are equipped with sharp claws. On the hind feet toes 2 through 5 have large claws whereas the first toe claw is rudimentary and may even be lacking.

Both the upper and lower lips in the genera *Rhyncholestes* and *Lestoros* have loose, fleshy lateral flaps of skin. These rare mammals are seldom captured, resulting in a paucity of information about their habits and ecology.

Dusky caenolestid (*Caenolestes fuliginosus*)
Mammal Image Library

Long-nosed caenolestid
(*Rhyncholestes raphanurus*)
Mammal Image Library

2. Manatees

The manatees are very large, slow-moving aquatic mammals with flipper-like forelimbs and no visible hind limbs. Their tails are flattened horizontally and broadened to provide propulsion through vertical movements (as opposed to fish, which move their tails laterally). Their nostrils are equipped with valves and are located at the tip of the muzzle and their eyes are relatively small. In a manner similar to many birds and some reptiles, their eyes have a well-developed nictitating membrane, a semitransparent eyelid that protects their eyes while allowing them to see. Manatees can hear but have no external ears. Their skin is thick, tough, and almost hairless, although stiff bristles appear around the lips.

Manatees eat aquatic vegetation, primarily sea grasses. An individual may consume up to 100 lbs (45 kg) in a single day. The name for the taxonomic order Sirenia refers to the possibility that the mermaid myth was based on the sighting of these animals, although on close examination this mistaken identity is hard to understand. One Old World species, the dugong, has two pectoral mammae that appear somewhat humanlike when the animal raises up from the water to peer into the distance. However, neither the dugong nor the manatees suckle young in the human fashion, despite the tales.

Manatees are susceptible to mutilation by the propellers of speeding pleasure boats. Many individuals show the regularly spaced slashes on their backs, and dead manatees with fresh propeller wounds are all too frequently recovered in near-shore waters.

Three species belong to the family Trichechidae, one is found on the west coast of Africa, but two are found in South America: the Amazonian manatee (*Trichechus inunguis*), which lives in the Amazon basin of Brazil, Colombia, Ecuador, Guyana, and Peru, and the larger West Indian manatee (*Trichechus manatus*), which occurs from the Gulf of Mexico and the Caribbean to the mouth of the Amazon. The Amazonian manatee is smaller than the West Indian manatee and lacks nails on the dorsal sur-

West Indian manatee (*Trichechus manatus*)

face of the flippers. Amazonian manatees also have white markings on their chests, a useful field mark for differentiation from their northern relatives.

The derivation of the word *manatee* seems to originate in Africa, brought over by enslaved peoples who were relocated to the American tropics. The word for the West African manatee in the Mandinga language is variously mantí or mandí, and apparently eventually became the familiar manatee.

From the 1600s through the 1700s manatees were much sought after, both in the Caribbean and throughout Brazil, for their savory meat and their fat and useful hides. The blubber was rendered to make excellent cooking oil and the hides were used to make leather. Strips of the hides were braided to make the whips used to punish the slaves. Later the hides were used to make heavy-duty leather. Presumably, from early on many thousands of manatees were killed for commerce. From 1935 to the early

1960s 4,000 to 10,000 manatees were harvested annually in Brazil. Since those days, manatee populations have decreased drastically, due to both commercial hunting and the drainage of swamps. Presently, most countries prohibit the killing of manatees, yet poaching continues here and there. Their numbers are recovering slowly.

West Indian manatee (*Trichechus manatus*)

3. Armadillos

The armadillo (order Cingulata) exemplifies the strangeness of some South American mammals. Armadillos are represented by 21 species in the tropics and subtropics of the Americas but are found nowhere else in the world. (The pangolins of Africa and Asia appear superficially like armadillos, but they are not at all closely related.)

Armadillos are strictly terrestrial; they never climb trees. Their diet, for the most part, is animal in nature; they eat insects and carrion, but also include succulent fruits and starch-rich plant tubers.

Armadillos are easily recognized because of their armor plating, which covers practically the entire body, except for the belly. The body is covered with a shell that has moveable plates or bands, permitting the animal

Llanos armadillo (*Dasypus sabanicola*)

Andean hairy armadillo (*Chaetophractus nationi*)

Big hairy armadillo (*Chaetophractus villosus*)

to bend. The top of the head and legs are covered with tough scutes and in some species large scales also cover the tail. This armor protects them from most predators. Two species of the genus *Tolypeutes* are capable of rolling up to form a tight ball, with no soft parts exposed. The armor plating is leatherlike in appearance but actually covers hard, bony scales.

Most species of armadillos have very little fur or hair. (The fairy armadillos are an exception.) Because of their physiology, they fail to maintain a constant high body temperature. Their body temperatures are lower than most mammals, 89.6°F to 96.8°F (32°C to 36°C).

Whereas the hind feet have five toes, the forefeet have from three to five toes according to species. Their legs are short and powerful and their toes are equipped with powerful claws. They are good diggers, excavating their own burrows. Once, a southern three-banded armadillo (*Tolypeutes matacus*), placed in a wire cage in a back yard for overnight keeping by a researcher, escaped by breaking through the wire cage floor and then burrowing a tunnel to freedom. And, although they are terrestrial, armadillos can swim well.

The story of leprosy and armadillos is a fascinating tale. Because *Microbacterium leprae,* the bacterium that causes the disease, grows best in colder tissues of humans (such as the tips of fingers and toes), researchers tried to grow it in naturally cold armadillos. The results were positive in the armadillo genus *Dasypus* and scientists had hopes of using armadillos to produce a vaccine against leprosy. Successful vaccines were actually developed, but medications to treat leprosy were also developed that to many seemed more effective. Even though interest in the vaccine has waned, armadillos are still considered a model for studying the bacterium, which has proven impossible to grow in the lab.

Leprosy was subsequently discovered in natural populations of armadillos, leading some to question whether animals infected by researchers had escaped. This idea has, however, been dismissed for two reasons. First, it does not appear that any armadillos infected with *Microbacterium leprae* ever escaped. Second, the two laboratories that conducted the studies, one near Baton Rouge, Louisiana, and the other on the Atlantic coast of Florida were just too far from the various locations where naturally occurring leprosy was found (Mississippi, Texas, and Mexico), given the time between when the studies took place and when leprosy was discovered in natural populations.

In South America, the folklore in several countries alleges that eating armadillos can cause leprosy. Presently it is not known how leprosy is transmitted, neither among humans nor in armadillos, but it is considered a contagious disease, so handling armadillos should be done with sanitary precautions.

Several species of armadillos belonging to a variety of genera are found in the zoos of North and South America, but the genus *Dasypus* is seldom if ever seen in zoos. The reason is their diet. *Dasypus* (commonly called the "long-nosed armadillos") is much more stringently insectivo-

Yellow armadillo (*Euphractus sexcinctus*)

Pichi (*Zaedyus pichiy*)

rous than the other genera and does not consume vegetable matter. The genera *Chaetophractus, Euphractus,* and *Zaedyus,* on the other hand, will accept food such as carrots, potatoes, and meat. During leprosy studies, a novel approach was used to keep captive *Dasypus.* The armadillos were fed cat food that had chopped-up earthworms added. The armadillos avidly sought the earthworms and incidentally consumed some cat food. Subsequently, fewer and fewer earthworms were included in the cat food, until the armadillos had adapted to eating pure cat food.

Southern three-banded armadillo (*Tolypeutes matacus*)

Brazilian three-banded armadillo (*Tolypeutes tricinctus*)

4. Sloths and Anteaters

The strange behavior of sloths, hanging from the undersides of tree branches, with motions that approach the movements of the hands of a clock, has fascinated naturalists for centuries. These almost unreal animals are found only in the New World tropics and belong to two groups in the taxonomic suborder Folivora. The three-toed sloths are members of the family Bradypodidae, which has a single genus, *Bradypus*. The two-toed sloths belong to the family Megalonichidae, which also has a single genus, *Choloepus*.

The hair of sloths is different in form and structure from that of other mammals in two respects. First, their hair has microscopic grooves that allow for the attachment and growth of blue-green algae. The algae impart a greenish color to their fur, providing camouflage while in the tree canopy. Second, their hair flows from the center of the stomach to the back, allowing the rain to shed off the belly while the animal hangs upside-down. For most mammals, the hair flows from the back to the center of the stomach.

The structure of the sloths re-curved claws and the articulations of their toes permit them to hang and even sleep in the upside-down position. They also have a low center of gravity, which, in combination with their strong grip, diminishes their need for agility.

The sloth's diet is composed mainly of tree leaves, which while abundant and easily available, have low nutrient value and are difficult to digest. Digestion is aided by symbiotic bacteria harbored in a many-chambered stomach. These bacteria can perforate the cellulose of the cell walls, thus making available the nutritionally valuable cell contents. Still, the amount of energy consumed is too little to sustain most animals. In addition to adjusting their pace-of-life to match low-energy intake, sloth metabolism is low, as is their body temperature (as low as 80.6°F [27°C]).

Three-toed sloth
(*Bradypus variegatus*)

Sloths may be important hosts for several important arboviruses (arthropod-borne viruses) such as yellow fever, St Louis encephalitis, Ilheus virus, and Venezuelan encephalitis. Sloths may have the virus circulating in their blood for long periods, and being relatively large, may infect many mosquitoes, which then may transmit the disease to humans or other animals. Two-toed sloths are also hosts for *Leishmania brasiliensis panamensis,* which causes cutaneous leishmaniasis in humans in Panama.

Little information exists on the relationship between the ancient inhabitants of South America and sloths or anteaters, but some anthropologists have suggested that early human inhabitants killed the last of the giant ground sloths, because those species became extinct some 10,000 years ago. Today's sloths are an important food item, but for Harpy Eagles, the largest of the birds of prey in the Amazon rain forest. The eagles aside, sloths apparently have a low mortality rate as judged by their very low reproductive rate of one young per year per female. Curiously, sloths find it inappropriate to relieve themselves while hanging from branches and, consequently once or twice a week, they descend to the ground to urinate and defecate. Sloths tolerate humans well and are frequently permanent residents in the trees of downtown plazas of major towns and cities in tropical America.

In comparison with sloths, the arboreal anteaters have a high center of gravity and use their extreme agility to keep from falling. The anteaters (families Myrmecophagidae and Cyclopedidae) are also called vermilinguas in some circles, a reference to their long, thin tongues. They are perhaps the strangest of all South American mammals, with their slender long, tubelike snouts. They appear to be walking on their knuckles. And,

Two-toed sloth
(*Choloepus didactylus*)

Giant anteater (*Myrmecophaga tridactyla*)

in the case of the giant anteater, striking stripes on the face and chest break up the outline of the animal when it stands at the edge of a forest.

There is a wide disparity in size between the three genera of anteaters, *Cyclopes, Tamandua,* and *Myrmecophaga.* The smallest, the silky anteater (*Cyclopes didactylus*), weighs 5.1 to 9.2 oz (145 to 260 g), the medium tamanduas (*Tamandua mexicana* and *T. tetradactyla*) weigh 9.9 to 18.5 lbs (3.6 to 8.4 kg), and the giant anteater (*Myrmecophaga tridactyla*) weighs 48.5 to 86 lbs (22 to 39 kg).

Although the species name of the giant anteater, *tridactyla,* would indicate three toes, in fact there are four. The fourth toe is reduced greatly in size, however, whereas the other three contain rather large claws. Another etymological puzzle involves the tamanduas, whose scientific name, *Tamandua tetradactyla,* indicates that it has four toes. However, the hind feet of this species have five toes each, all of which have small claws.

The tails of the tree-climbing anteaters (the silky anteaters and the tamanduas) are prehensile, allowing them to grasp branches. Fur from the body extends onto the base of the tail, but the terminal part, especially beneath, is without hair, allowing it to grip branches. In contrast, the tail of the giant anteater is neither prehensile nor bare. In fact it is covered with very long hairs directed vertically both above and below. The hair is laterally compressed, giving the appearance of a furred palm frond or flag. Giant anteaters cover themselves with their tails when sleeping. Usually they sleep beneath the shade of small trees or bushes, but occasionally they may sleep in the middle of a field. Although it appears that they would be readily visible, in fact, their coloration, in particular, of the tail that covers them, makes them resemble a lump of grass.

Anteaters are active both by day and night, with slight variations in diet among the species. In nature, the diet of the giant anteater is ants, termites, and beetle larvae. After feeding sparsely on a colony they move on to other opportunities, thus leaving their prey relatively undamaged for a return visit in the future. The tamanduas also feed on ants and termites, both terrestrial and tree termites, as well as bees. The silky anteaters feed primarily on ants but also consume some tree termites. They use their strong claws to tear open ant and termite nests, after which they extend their long, sticky tongues into the nests and retract them covered with their prey. Tamanduas are widespread in many parts of South America and are one of the animals most commonly seen dead along the roadside.

The anteaters are totally lacking in teeth. The muzzle is extremely thin, elongated, and downwardly curved, far more than any other mammal. The tongue has tiny spines directed backward and covered with

Sleeping giant anteater

Silky anteater (*Cyclopes didactylus*)
Gloria Jafet, São Paulo Zoo

Southern tamandua (*Tamandua tetradactyla*) (Venezuela)

Southern tamandua (*Tamandua tetradactyla*) (Brazil)

Giant anteater (*Myrmecophaga tridactyla*)

sticky saliva to aid in feeding. The claws on the forefeet are long and sharp for ripping open the nests of ants and termites. When threatened, anteaters may use their claws for defense as they rise up on their hind feet and face the enemy. Giant anteaters are said to have disemboweled dogs and are fully capable of defending themselves against jaguars and pumas.

5. Monkeys

The spectacular monkeys of South America belong to the group called the Platyrrhini, which are distinguished from the Old World monkeys and apes of the suborder Catarrhini. The principal difference in appearance is the placement of the nostrils, which are directed laterally in the New World species and downward (as in humans) in the Old World species. None of the Old World species have a prehensile tail, but several of the New World species do. All of the New World monkeys are primarily arboreal, while some of the Old World species are basically terrestrial.

Traditionally, the New World (or American) monkeys were divided into two families, the Cebidae and the Callitrichidae. The family name Cebidae remains and contains three subfamilies, the Callitrichinae (marmosets and tamarins), the Cebinae (capuchins), and the Saimiriinae (squirrel monkeys). The family Callitrichidae has recently been subdivided into three families: the family Aotidae (night monkeys); the family Pitheciidae, with two subfamilies, the Callicebinae (titis) and the Pitheciinae (sakis and uacaris); and the family Atelidae, with two subfamilies, the Alouattinae (howler monkeys) and Atelinae (spider, woolly spider, and woolly monkeys).

The diversity of the various monkeys requires further exploration. Beginning with the marmosets and tamarins, we find fairly small monkeys with long nonprehensile tails. The fur is soft and dense or silky. Some have adornments on the head of ear tufts, moustaches, ruffs, or manes. The face is either naked or sparsely haired. The forelimbs are shorter than the hind limbs; the thumbs of the forefeet (hands) are nonopposable, but the large toes of the hind feet are opposable. The toes have claws, not nails as in other primates, except for the large toe of the hind foot, which has a nail. They feed on fruit, insects, and exudates of tree sap of which they may cause the flow by gouging holes in the bark. Marmosets are equipped with special teeth, enlarged lower incisors, with which they can gouge the bark of trees to cause the flow of exudates. This specialization permits marmosets to ensure a regular diet of carbohydrate, even when their favorite source, fruits, becomes scarce.

With one exception, marmosets all belong to the genus *Callithrix,* including the pigmy marmoset, the smallest of all monkeys. The exception is Goeldi's monkey (*Callimico goeldii*), a species that continues to evoke debate among taxonomists. It shares some features with the other Cebinae, but the structure of its molars differs from the family and unlike all the other marmosets and tamarins, which give birth to twins, Goeldi's monkeys usually give birth to single offspring.

The social structure of marmoset groups includes a physiological adaptation by which only the dominant female breeds. The subordinate marmoset females experience a suppression of ovulation, thereby ensuring that the dominant female will be the sole pregnant member of the group. The rationale for marmosets having only one female give birth is complex. Perhaps it is useful to have all the other females help in rearing

Red howler monkey (*Alouatta seniculus*)

Goeldi's monkey (*Callimico goeldii*)

Geoffroy's marmoset (*Callithrix geoffroyi*)

the twins. Additionally it results in the nonbreeding females gaining experience in taking care of infants, which will be useful when it is their turn to raise their own offspring.

Tamarins, a favorite attraction at most zoos because of their active lifestyle, belong to either of two genera, *Saguinus* or *Leontopithecus*. Tamarins feed opportunistically on gum exudates, except that the saddle-backed tamarin (*Saguinus fuscicollis*) will take advantage of gum-feeding holes made by the pigmy marmoset. Tamarins are described as being the most generalized of living primates.

White tufted-ear marmoset
(*Callithrix jacchus*)

Pigmy marmoset
(*Callithrix pygmaea*)

In tamarins there is no suppression of ovulation as in the marmosets, yet a type of social control occurs. Ovarian cycles are synchronized such that all the members of the group ovulate together. The dominant female monopolizes the males, however, at the time that all are simultaneously fertile. Nevertheless several female tamarins do become pregnant. Perhaps the dominant female depends less on the help of others to raise her young than the marmoset dominant female, but whatever the case, tamarin groups raise several young at the same time.

Golden-rumped lion tamarin (*Leontopithecus chrysopygus*)

Golden-headed lion tamarin
(*Leontopithecus chrysomelas*)

Golden lion tamarin (*Leontopithecus rosalia*)

Midas tamarin
(*Saguinus midas*)

Saddle-backed tamarin
(*Saguinus fuscicollis*)

Cotton-top tamarin
(*Saguinus oedipus*)

Golden-mantled tamarin (*Saguinus tripartitus*)

Capuchin monkeys are larger by far than the marmosets and tamarins. In contrast to the marmosets and tamarins, which have a total of eight molar teeth, the Cebid monkeys have a total of 12. Some species have prehensile tails and all are extremely agile in the trees. The eyes of these monkeys are placed looking foreword, providing stereoscopic vision, which provides superior depth perception, important to arboreal species that must accurately judge their jumps between branches. They also have good color vision.

Capuchins are omnivorous; they feed by day and eat fruit, insects, and leaves. The fruit-eating habits of most Capuchins make them significant dispersers of seeds of tropical trees, and, similar to bats, they are important to maintenance of the tropical forests. They drink water by sucking.

Only very rarely do they descend to the ground; they are primarily arboreal and as such are seldom found in open savannas. Nevertheless, some monkeys will cross open areas to reach isolated patches of forests. Capuchins are commonly kept as pets, and used in films and for medical research.

White-fronted capuchin
(*Cebus albifrons*)

Brown capuchin (*Cebus apella*)

White-faced capuchin
(*Cebus capucinus*)

Weeping capuchin (*Cebus olivaceus*)

Weeping capuchin
(*Cebus olivaceus*)

Squirrel monkeys are instantly recognizable for their distinctive faces, with a white mask, dark muzzle, and dark cap. Their long flexible but nonprehensile tails with dark tips are also distinctive. They are very social and travel in noisy groups, squealing, chirping and in general crashing about the tree limbs.

Night monkeys, also appropriately called owl monkeys, are the only nocturnal monkeys, and as such possess enormous eyes. The head is round and the ears are short, almost hidden in the fur. The tail is long, as long as the head and body, it is covered with fur and is not prehensile. Their fingers and toes are long and narrow and have expanded pads at the tips and large pads at their bases, which are furrowed with noticeable lines of papillae. Night monkeys are found in small groups of two to five, uttering soft, low calls and louder owl-like hoots. These small groups are, in fact, the adult pair, plus the baby of the year, the one-year-old and possibly the 2-year-old. By 3 years the young disperse, search out their respective mates, and form a new pair, usually in a new territory. They feed on fruits, young leaves, insects, nectar of flowers, and occasionally small vertebrates. Although they may come close to human camps and

Squirrel monkey (*Saimiri sciureus*)

their eyes shine brightly, they avoid direct beams and usually leave when discovered. Night monkeys' eyes are evolved from some diurnal ancestors. Thus, unlike other nocturnal mammals, they lack the *tapetum lucidum,* that reflective layer of the retina giving the green eye shine to many nocturnal species. However, similar to other nocturnal species such as owls, they have evolved large eyes with rounded lenses, which gather more light to focus on the retina. Additionally they have evolved a higher than usual density of rods compared with diurnal primates. The eyes of night monkeys have been used extensively for research because of their large size and the clarity with which it is possible to view their retinas. Because they are adapted for night vision, their iris can dilate to nearly the diameter of the eye at its equator, permitting a full view of the fundus of the eye.

Night monkeys have been used by researchers to improve antimalarial drugs. While they do not appear to be frequently infected naturally with the human malarial parasites, they nevertheless can be adapted to serve as hosts, especially if their spleens are removed. Their use in this regard has contributed significantly to advancing more effective treatment of malaria in humans.

Azara's night monkey (*Aotus azarae*)

Humboldt's night monkey
(*Aotus trivirgatus*)

Black-headed night monkey (*Aotus nigriceps*)

The family Pitheciidae contains the titi monkeys (genus *Callicebus*) with about 28 or more species, the uakari monkeys (genus *Cacajao*) with two species, the bearded saki monkeys (genus *Chiripotes*) with four species, and the saki monkeys (genus *Pithecia*) with five species. All have nonprehensile tails and in the *Cacajao* the tails are shortened and stumpy.

Many consider the monkeys of the subfamily Callicebinae, commonly called titis, to be among the most attractive monkeys. With their beautiful, harmonious colors, long, thick and in some cases almost silky fur, they are without doubt very pretty animals.

Dusky titi (*Callicebus moloch*)

The sakis, bearded sakis, and uacaris are a subfamily (Pitheciinae) of three genera and nine species. As a group, they are notable for having abundant and dense hair over most of the body except the face and hands. The bearded sakis have a shock of hair on the top of their heads, parted in the middle, which appears like a wig. Other sakis have a conspicuous hood of long, coarse hair directed forward, which partly conceals the ears and face. For the sakis, the tail is long and bushy, giving rise to the term "fox-tailed monkeys." For the uacaris the tail is shortened. They are the only short-tailed American monkeys. The face of the uacaris is nearly without hairs; and in one species is colored a striking red or pink, the other is black skinned, and their ears have the shape of human ears. The males of one species of saki have a strikingly white-haired face. The beards of the bearded sakis are humanlike.

Equatorial saki (*Pithecia aequatorialis*)

Monk saki (*Pithecia monachus*)

Pale-faced saki
(*Pithecia pithecia*)

The family Atelidae comprises the howler monkeys, spider monkeys, woolly monkeys, and the woolly spider monkey. The prehensile tails are naked on the underside at the terminal end, providing a better grasp of branches. While the howler monkeys have well-developed thumbs on their hands, the thumbs on the hands of spider monkeys are greatly reduced in size or simply lacking.

Howler monkeys are most remarkable for their sonorous voice. Their voice has been likened in volume to that of the roar of a lion. An enlargement of the hyoid bone and the lower jaw facilitates this remarkable voice. Howler monkeys may call throughout the day and sometimes by night, but they tend to carry on a resounding chorus at dawn and dusk. The presumption as to why they call so much is that social groups keep track of each other, thus allowing neighboring groups (whose home ranges often overlap) to avoid each other. Such avoidance reduces territorial fights.

Howlers are quite agile in passing from tree to tree, and, employing their prehensile tails to considerable advantage, they achieve remarkable leaps. While seen most often resting in riverside trees, they nevertheless can also populate and survive as social groups for years in relatively small clumps of isolated trees in the open Llanos. Howler monkeys eat large amounts of leaves, usually the fresh green young leaves, and also consume fruits, buds, and even flowers.

Female black howler monkey
(*Alouatta caraya*)

Mantled howler monkey (*Alouatta palliata*)

Red howler monkey (*Alouatta seniculus*)

Red howler monkey (*Alouatta seniculus*)

The spider monkeys, woolly monkeys, and the woolly spider monkey constitute the subfamily Atelinae. All are relatively large, but not heavy bodied like the howler monkeys. In *Ateles* and *Brachyteles* the thumbs (of the hands) are rudimentary or vestigial, but the large toes of the hind feet are opposable and are used for grasping. The spider monkeys are without doubt the most agile of all South American monkeys. Their arms and legs are exceptionally long, as is their superior prehensile tail. While other monkeys use their prehensile tail to aid their movements in the trees, spider monkeys appear to depend primarily on their tail in traveling, with their four limbs assisting in the process. Their speed and agility in the trees is remarkable and appears to depend on possessing a visual capability in rapid accommodation (focusing) beyond human capacities. The clitoris of females is quite large and often mistaken for a penis. As a result, there were indigenous tribes that believed some spider monkey groups were composed of males only.

White-bellied spider monkey
(*Ateles belzebuth*)

Yellow fever is a viral disease of great historical repute. It has two epidemiological variants, the urban cycle and jungle yellow fever. Monkeys are the natural host of jungle yellow fever, which is frequently found in the tropical forests of South America. Based on those species already known to be involved, night monkeys (*Aotus* sp.), howler monkeys (*Alouatta* sp.), capuchin monkeys (*Cebus* sp.), spider monkeys (*Ateles* sp.), squir-

Black spider monkey (*Ateles paniscus*)

Brown woolly monkey (*Lagothrix lagotricha*)

rel monkeys (*Saimiri* sp.), and marmosets (*Callithrix* sp.), it is safe to say that all South American monkeys can be involved. Howler monkeys are highly susceptible and many die during outbreaks, while capuchin monkeys appear to be more resistant and thus may serve as reservoir hosts. Humans may occasionally be infected when they enter a forest where an outbreak is in progress.

6. Rodents

We mistakenly think of rodents as despicable animals. Our human prejudices against this group are based on knowledge formed from information gathered about the three species of rodents that have lived with us since the beginning of civilization. These species are the brown rat (*Rattus norvegicus*), roof rat (*Rattus rattus*), and the house mouse (*Mus musculus*). They are implicated in different diseases and even plagues that have affected us, so our attitude is understandable. Additionally, the brown rat, in particular, has adapted to living in man-made environments that are disgusting, our sewers.

Consequently, we fail to place this group in the high category that it deserves. Rodents excel both ecologically and taxonomically. Populations of rodents outnumber most other groups in most ecosystems. Their biomass (the combined weight of living animals), despite smaller size, but because of greater numbers, exceeds that of those species that prey on them. They are usually the principal item in the diet of the attractive predators we so admire. Without rodents, these predators could not survive.

Taxonomically, rodents comprise, by far, the largest group of mammals in South America. There are 16 families, 119 genera, and 497 species (nearly 50% of all South American mammals). One group, the infraorder Hystricognathi, differs from the family Cricetidae in several ways. Internally, many characteristics of the skull and teeth set them apart. Externally, many species differ visibly in the shape of their ears. Ordinary rodents (Cricetidae) have round ears, whereas many hystricognathous rodents have an indentation in the lateral edge of the ear.

An outstanding characteristic of rodents about which children learn early on from the comics is their prominent upper incisor teeth, made all the more outstanding by the absence of the canine teeth and a pronounced space (diastema) between the incisors and the premolars and molars. As gnawing animals, their prominent incisors aid them in this spe-

Bicolored arboreal rice rat
(*Oecomys bicolor*)

cialty. These incisors have a hard outside enamel, while the rest of the tooth is slightly softer, resulting in constant natural sharpening through gnawing as the softer dentine and internal enamel wear more quickly. This constant wear is balanced by the constant growth of the incisors.

Rodents have diverged greatly in their occupation of different habitats. Most are generalists, living on the surface of the ground and utilizing burrows for escape from predators. Nevertheless, many are highly arboreal, almost never descending to the ground, and many others have adapted to semiaquatic lives and have become excellent swimmers. Yet another group is almost totally subterranean. For the most part they are herbivorous, although most include animal protein in their diet, usually in the form of insects or carrion. Some aquatic forms have even specialized in eating crabs. The herbivorous forms seem to have split into specialists, with a minority becoming seedeaters and the majority consuming foliage or roots. From the point of view of ecological pyramids, rodents are the primary consumers and as such form the base of the pyramid, whereas the predators that eat them are the secondary and tertiary consumers, forming the upper, more narrow tiers.

Squirrels are well known and widespread throughout the world. In South America all members of this family are tree squirrels, but in North America and elsewhere there are ground squirrels, flying squirrels, chipmunks, and even marmots, besides the tree squirrels. Squirrels are seldom confused with other rodents. Their prominent bushy tails and tree-climbing habits set them apart. Squirrels even have the somewhat unique ability to reverse the position of their hind feet so they can cling to the trunk of a tree head down. All South American squirrels are active in the branches by day, descending to the ground mostly to travel to distant isolated trees.

The diet of squirrels is primarily nuts, fruit buds, and leaves, but may occasionally include insects, baby birds, and birds' eggs. The characteristic tail aids their agility in the trees, and it is used while sleeping to cover their feet and faces. Occasionally many melanistic (black) individuals arise within certain squirrel populations. Just as black gray squirrels are seen in parks in Washington, D.C., black squirrels also reside in some parks of the city of Caracas, Venezuela, but the species (*Sciurus granatensis*) is different, of course.

In South America there are 16 squirrel species belonging to only three genera. Curiously, all are confined to the tropical and subtropical regions, even though elsewhere, in the Northern Hemisphere, they abound in temperate and cold regions. Because they arrived from North America by way of the tropics of Central America, it is tempting to speculate that only tropical species made it and as yet they have not begun a re-evolu-

Red-tailed squirrel
(*Sciurus granatensis*)

Forest spiny pocket mouse
(*Heteromys anomalus*)

tion of temperate adaptability to occupy the cooler zones of the higher Andes Mountains.

In South America the spiny pocket mice of the family Heteromyidae are represented by one genus and three species, all distributed within the northwestern region of the continent. They are remarkable for possessing external fur-lined cheek pouches, a characteristic not found in other South American rodents. The mice use these cheek pouches to transport grains and nuts from wherever they find them to underground storage chambers in their system of tunnels. As such they sometimes are a nuisance to farms and granaries.

The family Muridae is the largest family of rodents in the world. Because of the accidental introduction of some species by humans in their travels in ships, this already nearly ubiquitous family has colonized practically the whole world. The subfamily found in South America is the introduced Murinae. Similar to the Cricetidae, they are distinguished from other South American rodents (the hystricognathous rodents) by their dental formula: incisors 1/1, canines 0/0, premolars 0/0, molars 3/3 x 2 = 16 (the hystricognathous rodents are incisors 1/1, canines 0/0, premolars 1/1, molars 3/3 x 2 = 20).

The subfamily Murinae is, in fact, an Old World family, but three species have been introduced accidentally to South America, as well as to just about the rest of the world. These species are the house mouse (*Mus musculus*), the brown rat (*Rattus norvegicus*), and the roof rat (*Rattus rattus*). They came first on the old sailing ships and continue arriving on the newer vessels. All three species are basically peridomestic and yet all three can also live far from human habitations. Nevertheless, they seldom thrive in natural habitats, instead they adapt to ecosystems disturbed by humans, such as air strips and abandoned fields, or corn and sugar cane fields.

House mouse (*Mus musculus*)

Brown rat (*Rattus norvegicus*)

Roof rat (*Rattus rattus*)

The family Cricetidae includes the New World rats and mice, as well as rodents from the Old World such as hamsters, gerbils, and lemmings. The South American cricetids are, for the most part, smaller than many of the hystricognathous rodents so characteristic of the continent. Although primarily terrestrial, many species in this family are good climbers, others are adept at burrowing, and still others are semiaquatic. None are truly fossorial. They are primarily herbivorous, feeding on vegetation, plant seeds, fruits, fungi, and shoots and roots, yet they often include insects in their diet and some have specialized in aquatic food, such as crustaceans and small fish. They inhabit all corners of the continent, occupying every conceivable habitat. They invaded South America during the late Oligocene or early Eocene, traveling across the Panamanian land bridge and rapidly dispersing and evolving species to fill all available niches.

Long-haired grass mouse
(*Abrothrix longipilis*) N. Bonino

Azara's grass mouse
(*Akodon azarae*)

Gray leaf-eared mouse (*Graomys griseoflavus*)

Marsh rat (*Holochilus sciureus*)

Pittier's crab-eating rat (*Ichthyomys pittieri*)

Narrow-footed bristly mouse
(*Neacomys tenuipes*)

Unicolored arboreal rice rat
(*Oecomys concolor*)

As the most numerous group, relatively small in size and widely spread, they are consequently an important item in the diet of many predators. Likewise, it is not surprising that they are important reservoirs for many diseases affecting humans and domestic livestock. They are the principal host for two particularly disturbing viral diseases, Argentine hemorrhagic fever (caused by Junin virus) and Bolivian hemorrhagic fever (caused by Machupo virus), that infect humans indirectly through ingestion or inhalation of the virus, which cricetid rodents shed in their feces and urine. Presently, four other related viruses (Guanarito, Pichindé, Sabiá, and Tacaribe) are from various parts of South America. When these viruses attack humans in well-populated areas, they attract attention, but for the most part they go undetected when they kill isolated individuals in remote regions. These are among the newly emerging dis-

Bicolored arboreal rice rat
(*Oecomys bicolor*)

Long-tailed pigmy rice rat
(*Oligoryzomys longicaudatus*)
N. Bonino

Above and right: Tome's rice rat (*Oryzomys albigularis*)

Coues's rice rat (*Oryzomys couesi*)

Big-headed rice rat
(*Oryzomys megalocephalus*)

Yellow-rumped leaf-eared mouse
(*Phyllotis xanthopygus*)
N. Bonino

Coues's climbing mouse
(*Rhipidomys couesi*)

eases that most likely will be making news as humans continue their invasion of wilderness areas. Cricetid rodents are also reservoir hosts for many other important diseases affecting humans in South America, such as Venezuelan encephalitis, leptospirosis, Chagas disease, leishmaniasis, trichinosis, and hidatidosis.

The subfamily Sigmodontinae (of the family Cricetidae) is the largest subfamily of rodents, and indeed of mammals in South America. The 65 genera and 296 species of Sigmodontinae have colonized practically every possible ecosystem by adapting in remarkable ways, such as evolving species that are semiaquatic, species that are highly arboreal, and species that are primarily fossorial, but the large majority of them are generalists, capable of swimming, climbing, and digging burrows. They are important food items for predators and have evolved a correspondingly high reproductive rate to compensate for their high mortality rate.

These rodents appear to have invaded South America from North America, possibly as early as the Miocene. The first forms were probably evolved as forest species, which later invaded grasslands and deserts, thus giving rise to new tribes, genera, and species. It is likely that more than one invasion of South America occurred as well as passage of South American forms back to North America.

Alston's cotton rat (*Sigmodon alstoni*)

Charming climbing mouse (*Rhipidomys venustus*)

Left and right: Hispid cotton rat
(*Sigmodon hirsutus*)

Short-tailed cane mouse (*Zygodontomys brevicauda*)

Butcher, Oldfield mouse
(*Thomasomys laniger*)

Hystricognathous Rodents

The name *hystricognathous* is Greek in origin, with the first part derived from the Greek word for porcupine (hustrikhos) and the second part from the Greek word for jaw (gnathos). The latter refers to a distinct flaring of the posterior part of the lower jaw. When captured, these rodents are much less apt to bite than rodents of the families Cricetidae and Muridae. Among the present day South American hystricognathous rodents are the porcupines, chinchillas, vizcachas, pacaranas, cavies, capybaras, agoutis, pacas, and spiny rats. On the basis of tails, these rodents can be conveniently separated into four groups. The first group is made up of those species with very long, sharp spines meant for serious defense—the porcupines. Second are the hystricognaths that have curved tails with long stiff bristlelike hairs toward the tips. These are in the families Chinchillidae, Octodontidae, and Abrocomidae. (The families Dinomyidae and Ctenomyidae are exceptions to this characterization; their tails are straight.) The third group comprises those without tails or with extremely rudimentary tails: the capybaras, cavies, maras, agoutis, and pacas. The fourth grouping, those with tails like rats, are the "spiny rat group," the family Echimyidae.

Porcupines are recognized by everyone, primarily because of their remarkable spiny bodies. Unlike their North American relatives, the ends of the South American porcupines' tails are bare of spines and are usefully

Brazilian porcupine (*Coendou prehensilis*)

Paca (*Cuniculus paca*)

prehensile. Considerable controversy exists among the specialists concerning the number of genera and species found in South America. Porcupines are arboreal, feeding on the bark, shoots, leaves, fruits, and flowers of trees.

The family Chinchillidae is restricted to southern South America and contains the chinchillas and vizcachas. There are three genera and six species, and they all have thick fur as is appropriate for the cold climate experienced by mountain species. Only the plains vizcacha (*Lagostomus maximus*) has left the Andes Mountains to colonize the steppes of the Pampas, Patagonia, Bolivia, Paraguay, and Brazil. All have large prominent vibrissae (whiskers), and with many hystricomorph rodents, the tail is curved and has long stiff hairs toward the tip. The genera *Chinchilla* and *Lagidium* have hind feet with four toes, but *Lagostomus* has only three toes. The hind legs are long and strong, with long hind feet, but the fore legs are short with short feet. Chinchillas and vizcachas often advance by hopping like rabbits.

The Chinchillidae seek refuge in their burrows, which often are under rocks or in crevices, except for the plains vizcacha, which digs elaborate burrows and then shares them with other rodents, and even snakes. All Chinchillidae are herbivorous. Chinchillas, prized for their furs, have been hunted to near extinction in the wild.

Bicolor-spined porcupine
(*Coendou bicolor*)

Paraguay hairy dwarf porcupine
(*Sphiggurus spinosus*)

Brazilian porcupine
(*Coendou prehensilis*)

Chinchilla
(*Chinchilla lanigera*)

Southern vizcacha
(*Lagidium viscacia*)

Plains vizcacha
(*Lagostomus maximus*)

The family Dinomyidae contains an unusual rodent, the pacarana. This species (the sole species of its family) is found from the highlands of Venezuela through Colombia, Ecuador, and Peru to Bolivia. The pacarana is a large rodent with white spots on the flanks similar to the Pacas, but it has short legs, and in body form looks like a large cavy with a tail. Pacaranas are ground dwelling and burrow for shelter, yet they are capable of climbing and resting in trees. Their food consists of fruits, leaves, and shoots. They are naturally nocturnal.

The cavies and Patagonian "hares," or maras (*Dolichotis*), are members of the family Caviidae. This family of 16 species is confined to South America. The guinea pig (*Cavia porcellus*) was domesticated by native Americans of the Andes region from Venezuela to Chile, and although it is distinct from wild species, it appears to be closest to *Cavia aperea*.

The cavies have short limbs but the maras have long rabbitlike legs. Cavies practice coprophagy similar to capybaras and rabbits. Coprophagy is a behavioral digestive process in which an enlarged cecum processes food with the aid of symbiotic microorganisms capable of cellulose digestion. The process creates coecotrophs—large soft pellets that are reingested. During coprophagy, cavies raise themselves up on all four feet and reach down directly between their legs. The coecotroph is directed forward and taken rapidly, while normal feces are directed backward and thus expelled.

Pacarana (*Dinomys branickii*)

Brazilian guinea pig (*Cavia aperea*)

Domestic guinea pig (*Cavia porcellus*)

Mara (*Dolichotis patagonum*)

The Capybara (subfamily Hydrochoerinae, *Hydrochoerus hydrochaeris*) is the largest rodent not only in South America, but also in the world. It is semiaquatic, requiring water to escape from enemies, to bathe and maintain its skin free of mites that cause mange, to keep cool during the heat of the day, and to drink. Much of their food likewise depends on water. They eat many grass species as well as some aquatic plants such as water hyacinths. Their forefeet have four digits and the hind feet have only three. They live in social groups maintained by a dominant male with his harem of adult females and their young. Capybaras are a possible reservoir for brucellosis. They likewise are heavily infected with *Trypanosoma evansi,* which is known to veterinarians as a disease of horses. Nevertheless, the unusually high infection rates, often more than 50%, in capybaras may indicate that they are the enzootic host for this trypanosome. They are susceptible to infestations of mites (*Eutrombicula alfreddugsi, E. batatas,* and *Blankaartia alleeni*), which can cause mange but are naturally controlled by bathing in mud. Mud bathing occurs in the morning, followed by drying in strong sunlight; the dried mud is then removed by bathing in water. Capybaras suffer from a disease called "Papera," which means goiter in Spanish, but capybaras do not lack iodine in their diet. Rather they become infected with a bacterium that is common to several species of rodents, *Streptobacillus moniliformis.* Their throats become grossly swollen, which on autopsy are found to be filled with pus. While the throat is the most frequently affected region, sometimes the scrotum is swollen and susceptible to physical damage. Most infected animals die.

Capybara (*Hydrochoerus hydrochaeris*)

Capybara (*Hydrochoerus hydrochaeris*)

Yellow-headed caracara on capybara, symbiosis

The agoutis and the acouchis constitute this tropical American family of 2 genera and 13 species, which are found from southern Mexico to southern Brazil. Similar to the capybara, they possess four toes on the forefeet and 3 on the hind feet. Their bodies are large and rounded, yet they are quite agile and quickly flee from predators, still they form a significant part of the diet of the larger cats. The body is long, the legs are thin, the head is large with a blunt snout, the eyes are large and the ears wide, and the tail is rudimentary or absent. For the most part they are nocturnal, yet occasionally they may be active by day. It is worthwhile to refer to those who first described their habits, such as Azara, as related by Cabrera and Yepes in 1960. To describe the relative tameness of this group, Azara related his experience with an "acutí" that arrived with its legs tied together. He untied it without any struggle by the animal, which

Azara's agouti
(*Dasyprocta azarae*)

Black agouti
(*Dasyprocta fuliginosa*)

immediately began grooming its face and ears with its forefeet. He then offered it uncooked root of manioc which, although it had been recently captured, the agouti accepted and ate, stretching its legs as an indication of satisfaction. After it escaped and was later recaptured, and even bitten by a dog, when it was returned to its corral, it ate as if it had not been hurt.

Brazilian agouti (*Dasyprocta leporina*)

Rufous agouti (*Dasyprocta punctata*)

Red acuchi (*Myoprocta acouchy*)

Paca (*Cuniculus paca*)

The pacas are similar to the agoutis but readily distinguished by the rows of white spots on their flanks. There is but one genus and only two species, the lowland paca (*Cuniculus paca*) and the highland paca (*Cuniculus taczanowskii*). Their forefeet have 4 toes and their hind feet have 3 large toes and two very small toes that do not touch the ground.

The burrowing Tuco-tucos belong to the family Ctenomyidae, which has one genus with numerous species distributed from Peru throughout southern South America, including Brazil, Paraguay, Bolivia, Argentina, and Chile. They are strictly fossorial and seldom come to the surface; consequently, they are seldom seen. Not surprisingly, they prefer regions of loose, sandy soil, yet they are often found elsewhere and, considering their fossorial habits, are surprisingly robust animals. They are the size of rats, with heavy bodies, large heads, short, thick necks, very small ears, short strong legs, toes with strong claws that are larger on the forefeet, and a short cylindrical tail. They have large feet with wide palms and with a fringe of comblike bristles on the sides. From these "combs" they get their name *Ctenos* = comb and *mys* = mouse. They have long, thick, fine fur. Their eyes are small.

Their diet is strictly vegetarian, consisting of tubers and the succulent parts of plants. They are quite vocal, calling from their underground tunnels to communicate with each other. The calls become louder in the afternoons and evenings when, at times, they emerge from their tunnels to sit up and look around. One may leave the burrow entrance to seek out fresh green grass to collect and bring back to its underground nest, while its mate remains at the mouth of the burrow. To some the calls sound like tuco-tuco, thus their common names. The burrows vary in depth from 30 to 50 cm but horizontally may be many meters long, sinuous and branching. The diameter of the burrow is 5–6 cm. Surface exits are plugged with loose dirt. They tend to live close to each other, without forming a true colony (excepting *Ctenomys sociabilis*), and such areas are considered dangerous to walk on, especially for horses. They are found in a wide diversity of climates and habitats, but always in uncultivated soil. One (*Ctenomys lewisi*) even has aquatic habits, with its burrows always close to streams and often filled with water. They seem to select a soil of certain compactness and dampness and do not venture into neighboring dryer soils.

The family Octodontidae (degus, vizcacha rats, and tunducos) is restricted in its distribution to southwestern South America, to Peru, Bolivia, Argentina, and Chile. There are 7 genera and 11 species, which vary in size from a large rat to a large mouse. The body is robust to fat. The head is large, the fur is usually long and thick, and the hind feet each have four well-developed toes, each with its sharp, curved claw and stiff bristles extending beyond the claw. The medium to large ears are rounded, thick, and covered with short hairs. The cheek teeth are simplified with the form of a figure 8 on their chewing surface (providing the origin of the family name). The tail can be short or long with coarse hair, short near the base and longer toward the tip, terminating in a distinctive tuft of hairs.

They are herbivorous, eating leaves, seeds, grasses, roots, bulbs, tubers, bark, etc. They seek refuge in rock crevices, rock piles, and holes and excavate their own complicated burrow systems where they may dwell as colonies. They store food in chambers in their underground galleries. They may climb small trees and bushes. They have two litters per

Degu (*Octodon degus*)

year, with lactation lasting for two months. When Europeans arrived to Chile, they found the natives consuming these rats, as well as the food stored in their galleries. When pressed, they too consumed these food sources.

The family Abrocomidae (Chinchillones or chinchilla rats), like the Octodontidae, is likewise restricted to southwestern South America, to Peru, Bolivia, Argentina, and Chile. The chinchilla rats have soft, thick, silvery dense fur, giving them their common name, but their ratlike heads and tails distinguish them. They are smaller than other hystricomorph rodents. The three middle toes of the hind foot have stiff hairs projecting over the nails, probably to help comb out parasites and remove dirt. The

Golden vizcacha rat (*Pipanacoctomys aureus*)
M. Mares

Plains vizcacha rat (*Tympanoctomys barrerae*)
M. Mares

soles of their feet are granular. Their ears are large and rounded. Some species have vomerine-like structures similar to teeth on the palate, which have a specialized function in feeding.

Chinchillones are colonial, living in burrows and taking advantage of rock crevices. Their diet is exclusively herbivorous and though terrestrial, they climb bushes and small trees. With fur having almost the softness of chinchillas, they are becoming rare because of hunting.

The spiny rats belong to a large family (Echmyidae) comprising 18 genera and 71 species. They are large with a ratlike form, yet, in fact, they are more closely related to the cavies, capybaras, porcupines, and agoutis. They are usually the most abundant mammals in rain forests. Their distribution is confined to the American tropics. Many, but not all species, have spiny or bristly fur, the spines being mixed with the body fur. The spines are flat and flexible and, unlike porcupine spines, they are not barbed. The head and body are typically ratlike in appearance, but the ears have a distinct scalloped trailing border. Their tails are made to break off easily as a method to provide escape from predators. Consequently, it is common to find animals with short stubs for tails. They have small litters of from one to three young, born fully furred and with their eyes open. They may have two or three litters per year. The spiny rats (*Proechimys*) are primarily terrestrial, but other genera (*Echimys, Dactylomys, Phyllomys,* and *Isothrix*) are highly arboreal. Their food consists of grasses, vegetation, fruits, and tubers. When trapped they are not aggressive and, if handled gently, usually do not bite. These are the most abundant ro-

Guaira spiny rat
(*Proechimys guairae*)

Coypu (*Myocastor coypus*)

dents of the tropical forests. Species of the genus *Proechimys* are considered to be important reservoir hosts for the virus of Venezuelan equine encephalitis (VEE).

The family Myocastoridae contains only one genus and species, *Myocastor coypus*. The nutrias or coypus are semiaquatic and originally native to southern South America, to southern Brazil, Uruguay, Paraguay, Bolivia, Argentina, and Chile. They have been introduced on purpose or accidentally to North America, Europe, and Japan, however, where they often become a problem. Their burrows can damage dikes and irrigation canals and they can devastate the vegetation in wetlands.

7. Rabbits and Hares

Rabbits and hares belong to the order Lagomorpha and the family Leporidae. Primarily species of the Northern Hemisphere, four lagomorph species are now wild in South America. Two species in the rabbit genus *Sylvilagus* have successfully invaded South America from the north, although they may have been aided by humans. One of the *Sylvilagus* species is the familiar eastern cottontail. In South America it is found in Venezuela, Colombia, and on several islands off the northern shore of Venezuela. Given the distances and currents, it seems unlikely that the rabbits swam to these islands, raising the possibility that early humans introduced them as either pets or a food source. It is also possible that the genus was transported to South America from North America in the same manner by indigenous peoples.

The second species of *Sylvilagus* is the tapeti, which ranges from Mexico and Central America south through Colombia, Venezuela, Ecuador, Peru, Bolivia, Paraguay, and northern Argentina to Brazil. This species occupies a great span of climates from tropical forest to Páramo.

Along with South America's two species of *Sylvilagus* rabbits, the continent also hosts two introduced European lagomorph species. The European hare (*Lepus europeus*) and the European rabbit (*Oryctolagus cuniculus*) are now residents of Argentina and Chile. The European hares (and hares in general) differ from rabbits by having precocial (fully developed) young. Young hares are born with their eyes open and are ready to flee danger, whereas baby rabbits are born with their eyes closed and are placed in a nest; they take weeks to reach the agility of newborn hares.

In general, rabbits are easily recognizable and as such they require little description. They are closely related to rodents, but there are many differences between the two mammal orders. For example, the inner side of the upper incisor teeth of a rabbit has vestigial teeth behind the larger functional incisors. These small teeth and many features of the skull are the primary characteristics that specialists use to separate rabbits from rodents.

While the more familiar domestic rabbits, which are descendents of the wild European rabbit, will burrow and form a warren for placing nests to give birth and raise their young, South American rabbits of the genus *Sylvilagus* never make such warrens. Instead the female selects an appropriate site to make her nest a day or two before giving birth. She digs a shallow cuplike depression that she lines with dry grasses. Later she lines this nest with fur plucked from her own belly and then delivers her young into this nest. After giving birth the female carefully covers the nest with the dried grasses, making it virtually impossible to detect.

The young rabbits are born with their eyes closed and without fur, but they develop rapidly and leave the nest about two weeks after birth. Mother rabbits do not visit the nest frequently, coming once or twice during each 24-hour period to provide milk for the babies. The timing of the visits appears to be random. This apparent inattentiveness may be a trait evolved to save the nestlings from predators, as more frequent visits might alert predators to the location of the nest. Young rabbits begin eating green succulent vegetation almost immediately after leaving the nest. However, they remain in the vicinity of the nest, hiding in dense vegetation by day, and coming out after dark to be nursed by their mother.

The reproductive output (fecundity) of rabbits is legendary, but the average litter size varies considerably between regions. In colder regions the litter size is greater than in the tropics. Female rabbits can, and often do, mate almost immediately after giving birth. Even though the gesta-

Cottontail rabbit (*Sylvilagus floridanus*)

European rabbit (*Oryctolagus cuniculus*)

N. Bonino

European hare (*Lepus europeus*)

N. Bonino

tion period is a brief (30 days), there is enough time to raise the preceding litter to independence, before making a new nest to give birth to the next litter. Depending on the geographic region, females may have up to several litters each year. Unfortunately for the rabbits, their mortality rates are also high.

Rabbits, despite their alertness and their speed and agility in fleeing from a danger, are nevertheless the object of desire for many predators. When resting, rabbits seek sites that provide heavy cover, such as dense thorn bushes, under which they make a semipermanent nest called a "form." Usually the form is in a slight depression and lined thinly with dried grasses. This resting site may be habitually used for several days before changing to a different site. Almost always alert, rabbits are constantly aware of the approach of potential predators, but they rely to a considerable extent on their camouflage and flee only as a last resort. In addition to predation, wild rabbits also succumb to diseases, which for the most part remain unstudied, with the exception of myxomatosis (a viral disease), tularemia (rabbit fever), and Rocky Mountain spotted fever. The last two diseases can affect humans.

8. Shrews

Shrews belong to the order Insectivora, and those in South America belong to only one family, Soricidae, and a single New World genus, *Cryptotis*. Like most insectivores, they are relatively small as mammals go. Shrews have five toes on each of all four feet, and most have short and velvetlike fur. Their ears and eyes are noticeably small and their tails are short, less than half the length of the head and body. Shrews tend to have a very high metabolism and require food frequently. They are terrestrial, fossorial (i.e., they burrow beneath the ground), and nocturnal, and they favor humid temperate climates. In South America these conditions can be found in the northwestern Andes Mountains and Venezuela, from Colombia to Ecuador and Peru where the ten or so very similar species of shrews in South America can be found (although some accounts suggest there may be shrews in the mountainous regions of coastal Venezuela).

South American shrews weigh about 0.25 oz (4 to 6 g) and are quite active in their search for grubs and other insect life. They have 30 teeth, which are not white, as in most mammals, but instead have reddish-brown points. Other North American shrews, *Blarina* and *Sorex,* have 32 teeth, also with reddish-brown tips.

Shrews can be very beneficial to plants because they consume large quantities of insects that feed on the plants. Shrews also serve as prey for other species including birds. In addition, they help to aerate the soil by serving as tiny tillers under the ground.

9. Bats

Of the order Chiroptera, bats are easily distinguished from all other mammals because they are the only mammals with the ability to fly (some other mammals can glide). The South American bats all belong to the suborder Microchiroptera, separating them from the Old World fruit bats, the Megachiroptera. As the name implies, the Microchiroptera are relatively small. The suborder also contains the champions of echolocation, including bats that can fly in total darkness, avoiding even the most filamentous obstacles, and finding and capturing their prey while both are in flight.

Lesser dog-like bat (*Peropteryx macrotis*)

In South America there are nine families of bats, the Emballonuridae, Noctilionidae, Mormoopidae, Phyllostomidae, Natalidae, Furipteridae, Thyropteridae, Vespertilionidae, and Molossidae. Of these nine families, six are found exclusively in the Americas and most are found in the Neotropics. Only three families, the Emballonuridae, Vespertilionidae, and Molossidae, are found worldwide. The most diverse family, the Phyllostomidae, or leaf-nosed bats, have a noticeable "arrowhead"-shaped vertical nose-leaf on the tip of their noses, except the vampire bats (of which there are three species) in which the nose-leaf is seen as a vestigial fold of skin.

Bats can live for quite some time in nature. Indeed it has been noted that bats live, on average, some 3.5 times as long as similarly sized mammals. The survivorship curve for vampire bats in northern Argentina indicates that the older bats are approaching 20 years of age, and marked bats of at least five species have been recaptured in nature after at least 30 years. It is possible to determine the age of vampire bats by counting annual growth lines in their teeth. The maximum age ranges from 15 to 21 years, but this does not reflect the age of the majority of the population, as only a very few vampire bats lived to this age. The average age of vampire bats varies in different colonies, but ranges from 3.04 years (Brazil) to 3.40 years (Argentina).

Frosted sac-winged bat
(*Saccopteryx canescens*)

Greater sac-winged bat
(*Saccopteryx bilineata*)

Lesser sac-winged bat
(*Saccopteryx leptura*)

Salvin's big-eyed bat
(*Chiroderma salvini*)

Common vampire bat
(*Desmodus rotundus*)

Many bats are insectivorous, but a variety of species of South American bats feed on fruit, pollen, and nectar. Other species feed on birds (*Vampyrum*), frogs (*Trachops*), fish (*Noctilio*), and even other bats (*Chrotopterus*).

Perhaps the strangest of all are the vampire bats, subsisting exclusively on the blood of larger vertebrates, mostly mammals and birds. These "sanguivores" do not actually "suck" blood, but rather shave out a divot of the skin and lap up the blood.

Among the insectivorous species there is extensive specialization. Many feed only on flying insects while others specialize on gleaning insects from plants. Those feeding on flying insects may specialize in high-flying insects, capturing them during rapid flight, while others feeding closer to the ground find their prey by patrolling open spaces on the edge of forests.

Greater false vampire (*Vampyrum spectrum*)

Fringe-lipped bat (*Trachops cirrhosus*)

Greater bulldog bat (*Noctilio leporinus*)

Woolly false vampire bat
(*Chrotopterus auritus*)

Lesser bulldog bat (*Noctilio albiventris*)

Harmless serotine
(*Eptesicus innoxius*)

Tiny yellow bat
(*Rhogeesa minutilla*)

Thomas's yellow bat
(*Rhogeesa io*)

Some bats are migratory, moving to keep abreast of available food sources. As fruit becomes available seasonally, fruit-eating bats arrive to take advantage of the abundance of their food. Insectivorous species of temperate zones may migrate to the tropics in search of food, while other species go into hibernation in caves to wait out the shortage of food.

All South American bats are considered to be nocturnal, yet some, such as the common mastiff bat (*Molossus molossus*), seem to prefer beginning their evening activity early, while there is still light in the sky. Another species, the proboscis bat (*Rhynchonycteris naso*), is active by day in the deep shade of humid forests and in shady sites along streams.

There seems to be a "coevolution" between tropical bats and many trees within tropical forests. This coevolution has resulted in interdependence. Bats pollinate and distribute the seeds of a large variety of tropical forest tree species, while the trees provide food for the bats. The tropical forest likely would be very different if the bats were removed.

Common tent-making bat
(*Uroderma bilobatum*)

Davis's tent-making bat (*Uroderma magnirostrum*)

Pallas's mastiff bat
(*Molossus molossus*)

Proboscis bat
(*Rhynchonycteris naso*)

Broad-toothed tailless bat
(*Anoura latidens*)

Geoffroy's tailless bat
(*Anoura geoffroyi*)

Godman's long-tongued bat
(*Choeroniscus godmani*)

Miller's long-tongued bat
(*Glossophaga longirostris*)

Pallas's long-tongued bat (*Glossophaga soricina*)

Curaçao long-nosed bat
(*Leptonycteris curasoae*)

In South America an astounding 212 of the 937 (22%) species of mammals are bats. This is second only to the rodents, which represent 49% of South American mammal species. However, bats are a huge component of humid tropical ecosystems in South America, where they comprise 40% of the total species.

Although many fossil bats are known, dating to the early Eocene (about 50 million years ago), all such bats had clearly evolved the ability to fly. There is truly little evidence about the earliest stages in the evolution of this group, that is, stages in the evolution that precede the ability to fly. Perhaps someday someone will find the fossil predecessors of bats, but at present we have only various competing hypotheses as to what lineage early bats can be traced. The first and most prevalent speculations presume that the bat ancestors were gliders, much as are "flying" squirrels today. Some have suggested that insectivorous predecessors of bats began leaping between branches of trees to capture flying insects, and their outspread legs developed membranes to aid them in leaping farther. Because most bats roost in caves, and many caves have mud deposits, there is hope that some fossil predecessors of bats may one day be found.

Tricolored bat (*Glyphonycteris silvestris*)

Orange-throated bat
(*Lampronycteris brachyotis*)

Common sword-nosed bat (*Lonchorhina aurita*)

Long-legged bat
(*Macrophyllum macrophyllum*)

White-throated round-eared bat (*Lophostoma silvicolum*)

Little big-eared bat (*Micronycteris megalotis*)

Most humans are "eye minded," that is, we perceive the world about us through our eyes. The bats of South America, of course, are much less dependent on their eyes. Their perceptual world is based largely on high-frequency sound. Their echolocation is an acoustic orientation system based on emitting a series of ticklike chirps in the range of 20 to 150 kHz and then receiving the echoes in a pair of exceptional ears. In the lower 20-kHz range the sounds can be heard by many people, but in the higher ranges they cannot be heard. These sound impulses are often frequency modulated (FM), that is, they shift their frequency from higher to lower. Yet other impulses are of a constant frequency. Many species mix these two types of sound pulses. What bats can perceive through their echolocation remains difficult for us to imagine.

The wide spectrum of items consumed by different bat species reflects, or is reflected by, their individual evolutions in form and behavior. Observers of the temperate zone are used to thinking of bats as insectivorous, and tropical regions are a spectacle each evening as the free-tailed bats fill the evening skies to the point of physically chasing the nighthawks from their feeding. Later, other species can be heard squabbling in the mango trees over feeding rights on some preferred fruits.

Tiny big-eared bat (*Micronycteris minuta*)

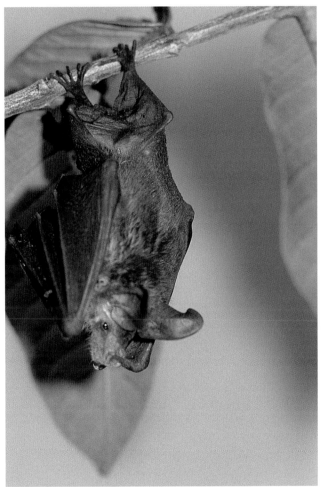

Schmidt's big-eared bat (*Micronycteris schmidtorum*)

Striped hairy-nosed bat
(*Mimon crenulatum*)

Pale spear-nosed bat
(*Phyllostomus discolor*)

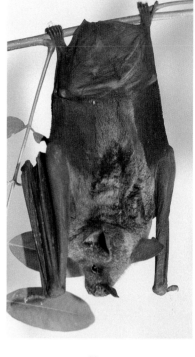

Lesser spear-nosed bat
(*Phyllostomus elongatus*)

Greater spear-nosed bat
(*Phyllostomus hastatus*)

Stripe-headed round-eared bat
(*Tonatia saurophila*)

Seba's short-tailed bat (*Carollia perspicillata*)

Silky short-tailed bat
(*Carollia brevicauda*)

The feeding preference of vampire bats is perhaps what most fasci-
nates people. Common vampire bats (*Desmodus rotundus*) are timid; they
avoid humans when they are awake, yet may prey on them if they are
asleep. These bats will roost in colonies numbering from 10 to 20 to as
many as 1,000 to 2,000. The colonies are segregated according to sex
into a principal colony of females, their nursing young, and a dominant
male who defends his harem from other males. Surrounding the principal
colony are satellite colonies of bachelor males. The principal colony occupies the best location in the cave or mine or the best hollow tree in the
region, and the satellite male colonies are relegated to other sites.

Little yellow-shouldered bat (*Sturnira lilium*)

Highland yellow-shouldered bat (*Sturnira ludovici*)

Gervais's fruit-eating bat
(*Artibeus cinereus*)

Vampire bats have been found in a very wide variety of habitats, some of them quite unusual. They reach their densest population when adequate roosts are found in reasonable proximity to domestic livestock, cattle, horses, pigs, or goats. Consequently, they are most often found in ranch lands or regions with many small farms. Less frequently, they may be found in humid tropical forests or along the seacoast. Yet exceptions occur. In Suriname vampire bats were responsible for cases of rabies in humans in an indigenous population located in humid tropical forest. Recently, in Colombia, Ecuador, and Peru predation by vampire bats on indigenous peoples has become a problem. In Venezuela vampire bats have been preying on fishermen and their families living on the coast of a small, off-shore island. In Chile vampire bats have adopted some seacoast caves from which they prey on seals.

Vampire bats, like most bats, are active only by night. They awaken from their sleep at about nightfall, but do not immediately leave the cave. Their first activities involve personal and social grooming, with much stretching of the wings and legs. The first bats to leave the cave are the bachelor males, who often attempt to enter the principal colony, but are immediately confronted by the dominant male. Usually they are repulsed and leave. An hour to two hours after dark the vampire bats begin leaving their caves and immediately go in search of food. They usually fly directly to the area where they fed successfully the previous night. If they find cattle resting in a pasture they visited the night before, they search for the same animal on which they fed previously. On finding this animal they proceed to the site where they fed previously and reopen the same wound. When a bat has had its fill, in about 10 minutes or less, it seeks a safe "night roost," some place where it can hang out of reach of predators, such as owls or bat falcons, and predigest their meal. The blood meal is mostly water and heavy, making flight laborious. At the night roost, under a bridge, overhanging cliff, or wide horizontal branch of a large tree, they begin urinating copiously, thus reducing their blood meal to a smaller more manageable clot, after which they return to their daytime roost. By two to three hours before day, all vampire bats have returned to their roost to sleep.

The principal food of the common vampire bat is the blood of mammals. The other two vampire bat species, the white-winged vampire (*Diaemus youngi*) and the hairy-legged vampire (*Diphylla ecaudata*) drink the blood of birds. The common vampire bat prefers the blood of bovines in most regions. Nevertheless, in some regions they show a clear preference for horses, even with abundant cattle nearby. In a laboratory experiment, a colony of adult vampire bats that had been captured from a region where they had been feeding on cattle was tested for their preference for blood offered in tubes. Results of two trials indicated that the blood most preferred was that of pigs, followed by human blood, and their traditional food, cattle blood, was third. The species least preferred by these common vampire bats were chicken and sheep. Other species, such as horses, foxes, etc., fell between the two extremes.

Modern human habitations form a partial barrier for vampire bats, but previously they fed extensively on Indians and European travelers who slept in hammocks without the cover of either tents or mosquito netting. Nevertheless, today, from time to time vampire bats revert to feeding on humans in their homes if windows are open and without metallic screening.

Vampire bats seek the neck, flanks, and rump in cattle, but another favorite site is just above the hoof, provided the cattle are standing on bare ground when they alight on the ground and approach them from the rear. In pigs, the ears and nipples are preferred biting sites, whereas in humans they seek the extremities, toes, elbows, nose, forehead, and shoulder blades.

Vampire bats travel an average of 9 miles (15 km) from their roosts to feed on cattle. However, if the cattle are located close to the roost they may spend time after feeding, exploring their environment. If the distance to the cattle is too far from the roost, they sooner or later must move their roost or change their selected hosts because the expenditure of energy in flying will exceed the intake of calories in food. As with other

White-winged vampire bat (*Diaemus youngi*)

Hairy-legged vampire bat (*Diphylla ecaudata*)

species, vampire bats may disperse widely from their birthplaces and take up residence in new roosts far away. Recapture of banded vampire bats made at a distance of about 60 miles (100 km) from the banding site is evidence of dispersal. Likewise, vampire bats visit neighboring colonies, which in most colonies results in a region being socially connected by the visits.

The nearly unlimited food supply for vampire bats in ranchlands means that population size is limited by roost availability. With roosts being limited to caves, tunnels, mines, abandoned buildings, and large hollow trees, the vampire bat population density is variable. In flatlands without caves, mines, or tunnels, vampire bat roosts are found along the courses of streams where large hollow trees grow, or where natural caves may sometimes be found in the banks of the rivers. Regions with limestone caves, areas with abandoned mines, and little-used railroad or irrigation tunnels are where the most dense vampire bat populations may be found. Nevertheless, there are exceptions, as in the Chaco Region of northwestern Argentina, where vampire bat roosts are located in large-diameter, deep water wells.

The gestation period for the common vampire bat is about 205 days, with only one young born (if twins are produced, only one survives). Lactation is variable, but may last several months. Most births coincide with the wet season, but births do occur year-round.

Blood, as a food, has many advantages. First, it does not need to be chewed. Blood contains all the proteins, carbohydrates, vitamins, and minerals needed by the host animal. The only disadvantages of blood as food are the excess of water and salt. Consequently, vampire bats have some of the best developed kidneys. But how vampire bats evolved their unusual diet remains an area of speculation. Some have suggested that the precursors of vampire bats were interested in the insects that were attracted to and even sometimes stuck in the natural wounds of larger animals. In the process of extracting these insects from the wounds, they sometimes consumed dried blood, and if they happened to reopen the wound, may have sampled fresh blood.

Because blood does not require mastication, vampire bats have vestigial molar teeth that cannot be used to chew effectively. Vampire bats have highly evolved specialized upper incisors, however, and also very special upper and lower canine teeth. Because their prey usually has fur, the canine teeth work as scissors to shave away the area where the bite will be made. Then the area is licked with saliva that contains both an anticoagulant and a local anesthetic. Finally the upper incisors open the wound, a nearly round shallow divot of skin. The bite is made by a quick downward movement of the upper incisors. After a brief interval, the "vampires" begin to lap up the blood that fills the wound. Grooves on both sides of the tongue work together with a groove in the middle of the lower lip to channel the flow of blood into the mouth. A vampire bat weighing slightly less than an ounce (25 g) may ingest up to 20 ml (0.67 oz) of blood. Pregnant females may feed twice in one night, exceeding the total of their body weight in blood consumed. In vampire bats, ingestion stimulates urination immediately. They often urinate on the animal on which they are feeding. Evidence indicates that this results in marking these individuals aiding return to them on subsequent nights.

After feeding, vampire bats usually retreat to a safe place, under an overhanging cliff or under a large horizontal branch, where they rest and excrete quantities of urine. Such sites are readily recognizable because the vampire bats also defecate, and the mixture of feces (made of broken red blood cells) and urine on the ground makes a characteristic tarlike mixture.

Vampire bats will tolerate having other vampire bats feed from the same bite. A new arrival may shoulder or back its way into the feeding bat, pushing it away from the bite, then begin feeding.

Vampire bats prefer to feed on moonless nights. They leave the roost about an hour or two after dark and often do not return until an hour or two before daylight. Vampire bats fly low, close to their prey, only a few feet above the ground or vegetation, similar to pilots "hedge-hopping."

Velvety fruit-eating bat
(*Enchisthenes hartii*)

Short-headed broad-nosed bat
(*Platyrrhinus brachycephalus*)

Hairy big-eyed bat (*Chiroderma villosum*)

Thomas's broad-nosed bat
(*Platyrrhinus dorsalis*)

Heller's broad-nosed bat (*Platyrrhinus helleri*)

Shadowy
broad-nosed bat
(*Platyrrhinus umbratus*)

Greater broad-nosed bat
(*Platyrrhinus vittatus*)

Female visored bat
(*Sphaeronycteris toxophyllum*)

Male visored bat
(*Sphaeronycteris toxophyllum*)

Jamaican fruit-eating bat
(*Artibeus jamaicensis*)

Between the useful insectivorous bats and the parasitic vampire bats lie the helpful fruit-eating species. Fruit-eating bats have some special problems with their food. Not only is it seasonal, requiring seeking out different trees or migration, but concentrations of bats on fruit-laden trees attract bat predators. Owls and bat falcons are perhaps the most successful, but arboreal snakes and mammals can also take advantage of bats attracted to the fruit trees. So, whenever possible, bats tear off the fruit in flight and carry it elsewhere to consume it in relative safety and solitude. This habit is so well entrenched in many bat species, that some tree species have evolved to employ bats for seed dispersal. The cashew nut tree (*Anacardium occidentale*) unlike most fruits, has its seed hanging down below the fruit. The fruit is sweet and delicious and bats tear ripe fruits away from the branches and fly to a nearby tree where they begin eating the soft upper parts. As opposed to the tasty fruit, the outer seed coating contains an irritating oil. Indeed, if the bats' lips make contact

Great fruit-eating bat (*Artibeus lituratus*)

with the seed, both the fruit and the seed are dropped in disgust. Thus, the seed is transported, neither consumed nor damaged.

We usually expect to find bats roosting in caves, mines, and tunnels during the day, yet their roosts are much more varied. Many mastiff bats, for example, prefer to rest under the traditional Spanish-type roof tiles. Large trees with dense foliage (e.g., *Ficus* sp.) are preferred by the ever-abundant common fruit-eating bat (*Artibeus jamaicensis*), which will also roost in caves. Yellow bats (*Lasiurus ega*) have even been seen roosting between boards stacked vertically and leaning against a garden wall. Attics often attract bats and years of guano accumulation can damage human dwellings.

Much less appreciated than the more familiar day roosts are the night roosts of bats. After their first feeding of the evening many species seek safe places to rest temporarily. These night roosts may be far from their habitual day roost, for example, under the eaves of porches, under bridges, overhanging cliffs, and large horizontal branches. The presence of bats can be detected through examining the floor or ground under the likely night roost for feces.

Southern yellow bat (*Lasiurus ega*)

Ghost-faced bat (*Mormoops megalophylla*)

Davy's naked-backed bat
(*Pteronotus davyi*)

Mexican funnel-eared bat
(*Natalus stramineus*)

Parnell's mustached bat (*Pteronotus parnellii*)

Trinidad funnel-eared bat
(*Natalus tumidirostris*)

Big bonneted bat
(*Eumops dabbnei*)

Wagner's bonneted bat
(*Eumops glaucinus*)

Dwarf dog-faced bat
(*Molossops temminckii*)

Bats are unusually agile creatures, both in flight and on the ground, in particular, in the region of their roosts. Each species flies in the air space for which it was evolved. We often see the agile twists and turns of insectivorous bats feeding in the light of street lamps. Children sometimes learn to capture them with butterfly nets, by tossing a small stone up into their flight path, then swooping the net from behind the bat, capturing both the bat and the small stone. An amazingly agile bat, the pigmy round-eared bat (*Lophostoma brasiliense*) appears to float like a butterfly in the space beneath the seat of a chair, inspecting leisurely where it will hang to rest. Unlike most species that fall to provide the momentum for flight, vampire bats can spring instantly into flight from the ground.

Pigmy round-eared bat (*Lophostoma brasiliense*)

Bats seem to bring about an adverse reaction in people who have had little contact with them. When questioned, these people often cite ideas about bats getting tangled in one's hair, that they harbor fleas or other parasites, and that they are carriers of rabies. In truth bats do not become tangled in people's hair, the parasites they harbor are specific for them and quickly abandon any humans on which they might happen to land, and although bats can transmit rabies, they are not "carriers" of rabies. Epidemiologically, a carrier transmits a disease over a considerable period of time without suffering from the disease. As mammals, bats are as much victims of the disease as humans are.

Black mastiff bat (*Molossus rufus*)

Bats can certainly transmit the rabies virus through bites, but the virus can also be spread via an aerosol route, for example, by breathing air in caves that contain large amounts of the virus.

Some bats (as well as other mammals) recover from rabies. When tested, many healthy individuals show rabies antibodies in their sera, indicating their past exposure. For example, some vampire bats die when exposed to the rabies virus, whereas others never develop symptoms, presumably because their immune systems destroy the virus. Still others of the population never become infected. Rabies outbreaks in vampire bats are transitory and migratory. The epidemic lasts for a few months and then moves on, just as it had arrived from some nearby infected population.

Broad-eared bat (*Nyctinomops laticaudatus*)

Brazilian free-tailed bat
(*Tadarida brasiliensis*)

Brown mastiff bat (*Promops nasutus*)

Brazilian brown bat (*Eptesicus brasiliensis*)

Diminutive serotine (*Eptesicus diminutus*)

Bats can be reservoirs for some other diseases that also affect humans or domestic animals. Perhaps the most outstanding of these diseases is histoplasmosis. Histoplasmosis is a respiratory disease caused by the fungus *Histoplasma capsulatum,* which has a yeast form in its parasitic phase. The guano of both birds and bats is the appropriate medium for the growth of this fungus. When the guano is disturbed, for example, by removing it either in cleaning an area or for use as fertilizer, spores are inhaled that can lead to serious infections, occasionally even to death. Persons who study bats in caves can, because of gradual exposure, develop immunity to histoplasmosis.

As we begin to appreciate the value of species diversity in the ecological scheme of things, we are beginning to ascribe greater importance to the place of bats. For example, tree fruits are evolved to enlist the aid of animals to disperse seeds, helping tree species in their quest through reproduction to maintain themselves in nature. Color was added to fruits to attract diurnal birds and monkeys, who are rewarded by food value as they disperse the tree's seeds. Besides the cashew tree previously mentioned, there are many tree species that have concentrated on bats to disperse their seeds. Such trees often do not add color to their fruits, it is their odor that attracts the bats, and their fruits may be green or dark brown. Pollination is primarily thought of as being carried out by insects, especially the bees, yet bats are such important pollinators that many species of tropical trees open their blooms at night, thus reserving them for bats or moths to pollinate. The bats are rewarded, of course, through receiving a nutritious drink of nectar. Also, some bat species consume the pollen itself. Pollen is messy food, of course. It sticks easily to the fur of the bats' faces and gets carried from flower to flower. Should one need reasons to appreciate bats, these are but a few.

Hoary bat (*Lasiurus cinereus*)

Argentine brown bat
(*Eptesicus furinalis*)

Big brown bat
(*Eptesicus fuscus*)

Silver-tipped myotis
(*Myotis albescens*)

Hairy-legged myotis
(*Myotis keaysi*)

Black myotis
(*Myotis nigricans*)

Chilean myotis
(*Myotis chiloensis*)
N. Bonino

10. Carnivores

We typically think of "lions, tigers, and bears, etc." when we hear the word "carnivore," but the name actually refers to the many species of the order Carnivora, a group of mammals that includes the dogs, cats, raccoons, weasels, skunks, seals, sea lions, bears, etc. All carnivores are adapted for killing and eating prey species, yet many are omnivorous, feeding on fruit and other vegetable material. The teeth are evolved to permit both killing and shearing flesh, but not for very much mastication of their food. Some are capable of climbing trees or swimming, but most are terrestrial. They can be as large as the jaguar and spectacled bear, or as small as a weasel.

The carnivores of South America are represented by 54 species in 9 families. The dog or canid family, Canidae, is composed of one wolf species—the aptly named "maned wolf" (*Chrysocyon brachyurus*)—and a wide variety of foxes. The large canine teeth possessed by members of the Canidae are used more to grasp prey than to kill, which is achieved through vigorous shaking. The prey species are usually much smaller, even for the very large maned wolf, which subsists principally on small rodents the size of the marsh rats of the genus *Holochilus*.

The cat family, Felidae, varies considerably in size, from the jaguar and puma to the smaller tigrillo and margay. In South America there are 12 species of wild cats, all of which are strictly carnivorous. Prey is captured through stealth and the use of long, retractable claws. Killing is most often accomplished by biting with the sharp, long canine teeth. The night vision of cats is well known, but they see just as well by day and their hunting is opportunistic, adapting to circumstances. The retina is equipped with a tapetum lucidum, a special reflective layer that is responsible for the bright eye shine of cats when caught in the glare of a flashlight. This layer provides a double stimulus to the rods, as the light strikes them both on entering and then again on exiting.

The claws of the feet are unusually sharp and important in the capture of prey. The claws are normally kept retracted in a special sheath, and consequently, unlike the foxes, cat's claws are not worn by walking on difficult terrain. They keep their claws strong and sharp by periodically scratching repeatedly a favorite tree trunk or limb. These scratch marks are striking and announce to the knowledgeable observer the presence of a puma or jaguar.

Jaguars are large enough to hunt humans and sometimes do, although they generally avoid them. The large puma also occasionally feeds on humans, but documented cases are rare. In general, all species seek prey much smaller than themselves and most humans are of a size that is formidable to these cats.

Jaguar
(*Panthera onca*)

In South America most of the native cats are spotted on the body and striped on the legs. Even the young of pumas are spotted, giving the impression that their ancestors were at one time also spotted. Even in the jaguarundi, some of the kittens are spotted. Unfortunately, a demand for the spotted skins of cats exists in the fur trade that has resulted in the illegal killing of many cats by poachers.

Cats are by nature solitary. They do not form social groups. In fact males and females come together only once each year when the female is in heat for the purpose of copulation. After repeated copulations with intermittent rests, the two then go their separate ways. Cats are at the top of the food chain. As such they depend on an abundant supply of prey. To survive they must kill at least several times each week, the frequency

Pampas cat
(*Leopardus colocolo*)

Geoffroy's cat
(*Leopardus geoffroyi*)

Ocelot (*Leopardus pardalis*)

Little spotted cat (*Leopardus tigrinus*)

Margay cat
(*Leopardus weidii*)

Jaguar
(*Panthera onca*)

depending on the size of the prey. Consequently, in natural systems it is not convenient to have several such predators in close proximity, as there might not be sufficient prey to support them all.

All of the South American canids have derived from North American forms. Their predecessors came across the Panamanian land bridge about 3 million years ago. The gray fox (*Urocyon cinereoargenteus*) is principally distributed in North America, but extends into northern Colombia and Venezuela. The short-eared dog (*Atelocynus microtis*) and the bush dog (*Speothos venaticus*) are derived from ancient forms of the genus *Cerdo-*

Puma (*Puma concolor*)

Jaguarundi (*Puma yagouaroundi*)

cyon to which the savanna fox (*Cerdocyn thous*) belongs. The various fox species of the genus *Lycalopex* are widespread and relatively abundant, feeding on a variety of foods, small mammals, birds, reptiles, as well as insects, carrion, fruit, and other vegetable material.

Considerably larger than the foxes, the maned wolf is a resident of the central portion of the continent, inhabiting Paraguay, Uruguay, the lowlands of Bolivia, the northeastern portions of Argentina, and parts of Brazil. Fossil forms of the maned wolf have been found in Mexico and Arizona, as well as many sites in South America, demonstrating that it once had a wider distribution.

With its reddish color and foxlike head and face, the maned wolf looks more like a very long-legged fox than a wolf. Likewise the maned wolves do not form packs, but hunt alone or at most in pairs. Only when young are accompanying females are they seen in larger groups. Their food, surprisingly, is mostly rodents and fruit. Seldom do they prey on anything larger. Both in appearance and habits maned wolves resemble the Abyssinian wolf of Ethiopia, to which it is not closely related. Maned wolves are rare and need protection. Their native habitat is rapidly being converted to agricultural purposes.

All fox species in South America are territorial in the sense that a pair, which tend to mate for life, defends the area in the center of their home range and near the den where the young are born and raised. Foxes in temperate zones mate toward the end of the winter. The gestation period is about two months, so the young are born in spring. The female

Maned wolf
(*Chrysocyon brachyurus*)
H. Delpietro

Savanna fox (*Cerdocyon thous*)

guards the young, even against her mate. The male brings food to the nursing mother and later, when the young can eat solid food, the male feeds them as well.

As they grow, the young are taken on trips into the field to learn how to find food. For several months thereafter these young occupy the home range of their parents. But as winter approaches, they begin to wander far afield and eventually seek out their own home ranges, distinct from that of their parents.

Foxes, unfortunately, frequently become infected with rabies virus. Outbreaks of rabies in foxes have occurred in widely separated parts of South America, such as the south end of Lake Maracaibo in Venezuela and the Province of La Pampa in Argentina. Outbreaks of rabies in foxes are seasonal and have two peaks, which are tied to the changes in habits of foxes during the year. The first (and larger) peak results from the mass dispersal of young foxes from the home ranges of their parents. As they wander through unfamiliar regions they are at the same time seeking a member of the opposite sex with which they will pair up for later mating. Consequently, there is considerable opportunity for infected foxes to meet with uninfected foxes, and thus an increased chance for virus transmission. The second and lower peak in fox rabies is tied to the mating season, when actual copulation occurs. Due to long incubation of the virus in foxes, disease peaks follow the dispersal of young and the mating season by about a month.

Culpeo fox
(*Lycalopex culpaeus*)

Patagonian gray fox
(*Lycalopex griseus*)

Pampas gray fox
(*Lycalopex gymnocercus*)

Bush dog
(*Speothos venaticus*)

Gray fox
(*Urocyon cinereoargenteus*)

The bear family (Ursidae) is widely found in Asia, Europe, North America, and South America. It is presently absent in Africa, it never occurred in Australia, and it would be absent in South America, except for the presence of a single species, the spectacled bear (*Tremarctos ornatus*). Named for its white facial markings, which on some individuals make it look like they are wearing glasses, the spectacled bears' markings are actually quite variable.

Spectacled bears live in the Andes Mountains, occupying a wide variety of habitats, from humid forests and scrub-thorn regions to high-alti-

Spectacled bear (*Tremarctos ornatus*)

tude grasslands. They feed on fruit, the bases of bromeliad leaves, the hearts of palms, bamboo hearts, corn, rodents, insects, and occasionally carrion. Spectacled bears are not only the sole bear in South America, but also the only member of the genus *Tremarctos*.

Members of the family Otariidae, or the "eared seals," are those seals that have an external ear. These seals also possess the ability to rotate their hind limbs forward, increasing their mobility on land. Eared seals are found on both the Atlantic and Pacific coasts of southern South America. Their diet consists of fish, squid, and small crustaceans, including krill.

Eared seals are highly gregarious during the breeding season. Males establish territories and defend them vigorously. When females arrive, males try to mate with as many as possible.

As opposed to the eared seals, members of the family Phocidae lack external ears and thus are called "the earless seals." They are readily distinguished from their eared cousins not only by the lack of external ears, but also by not having the ability to rotate their hind limbs forward. This makes their locomotion on land much more awkward. The earless seals use their hind limbs primarily for swimming, whereas the eared seals depend mostly on their forelimbs for aquatic locomotion.

Earless seals feed primarily on fish, shellfish, and squid. The famous leopard seal (*Hydrurga leptonyx*), however, earned its name from the fact

South American sea lion (*Otaria flavescens*)

Young of southern elephant seal (*Mirounga leonina*)

Southern elephant seal (*Mirounga leonina*)

Group of resting elephant seals
(*Mirounga leonina*)

that it hunts not only fish and squid, but also penguins and even other seals. The earless seals are not as gregarious as the eared seals, with most species of earless seals forming mating pairs. An exception is the male southern elephant seal (*Mirounga leonina*), which will establish territories and may accumulate a harem of up to 30 females.

The family Mustelidae in South America contains not only otters and weasels, but also grisons. Mustelids tend to have low-slung, long bodies supported by short legs. They have small heads, eyes, and ears, and well-developed tails that are shorter than the length of the head and body. All have powerful scent glands with two openings, one on each side of the anal aperture. Although the glands impart a strong odor, they do not squirt their scent as a form of defense as skunks do. The skull of mustelids is quite similar in appearance to that of skunks, with their short facial part, the very large cerebral space, somewhat flattened, and the very flat auditory bulbs.

The body form of mustelids permits them to have great agility and many are capable of unusual arboreal feats. They are known to be literally bloodthirsty, sometimes seemingly killing for that purpose only, as they then forego consumption of the meat. The native otters and North American mink (introduced to Chile) are exceptionally fine swimmers and can readily capture active fish.

The color of grisons is very curious. In the majority of mammals, and most other vertebrates (fish, amphibians, reptiles, and birds), the under parts are notably paler than the back and top of the head. In grisons the opposite is true, the throat and belly are deep, ebony black. As if to emphasize the difference, the sharp line of demarcation between the throat and the top of the head is exaggerated by a white border between the silver gray of the upper parts and the black of the throat. The advantage of this unusual coloration has escaped naturalists over the years and remains a mystery.

Grisons are social and have been reported to carry on unusual activities above and about the entrances to vizcacha (a groundhog-sized rodent, see chapter 6) burrows. A dozen or so grisons were observed in an involved and complicated dance, so involved in their activities that they failed to detect the arrival of the naturalist. They ran and jumped across the burrows crisscrossing each other without bumping or even touching one another, with a speed that confused the observer and made it impossible to follow the actions of a given individual. When the naturalist

Giant otter (*Pteronura brasiliensis*)

Tayra (*Eira barbara*)

Lesser grison (*Galictis cuja*)

Greater grison (*Galictis vittata*)

Patagonian weasel (*Lyncodon patagonicus*)
N. Bonino

moved in closer and was discovered, in alarm they dove into the burrows. Seconds later they were stretching their long black necks out of the burrow, growling and grinding their teeth and peering with their fiercely brilliant eyes.

Skunks, members of the family Mephitidae, were previously included as members of the mustelid family—the badgers, otters, etc. However, skunks have a different body form than the mustelids: they are more rounded and robust and have long luxuriant fur. Yet the skulls of skunks reveal their close affinity to the mustelids. Skunks are famously distinguished by their powerful scent glands located on either side of the anus. These glands can deliver well-directed, powerful squirts of scent directly into the face and eyes of a potential predator. Perhaps because they have little to fear from most terrestrial predators, skunks are relatively slow in their movements. However, Great Horned Owls do prey on skunks without discomfort.

Skunk coloration, usually a bright, contrasting combination of bold black and white body stripes, is an indicator to potential predators of something different. Young predators, lacking experience, may attempt to approach. They may be sprayed directly in the face and eyes, with the skunk simply raising its tail and directing the two jets with considerable accuracy. Even when facing its aggressor, skunks can do a handstand, bend over backward and deliver the jets forward into the face of their enemy. The result to the eyes is painful and temporarily blinding. The powerful odor is repugnant and causes much sneezing and sputtering on the part of the predator. All thought of attack is abandoned and the skunk walks away casually.

In South America the bulk of the family Procyonidae are ring-tailed species. Typical ring-tails include raccoons, coatis, and olingos. The kinkajous, which bear no rings on their tails, are also members of this family.

Members this family are highly omnivorous, eating animal material at times, but they are much more likely to consume fruit and other plant matter. All the procyonids are good climbers and several species spend much time in trees.

The South American raccoon is lankier than its North American relative, with a somewhat less distinct face mask, but with many of the same habits, even that of washing its food in water, for which it is called in some regions "osito lavador" (little washing bear). It is reasonably abundant and its tracks are to be seen in the mud on the banks of streams and even within caves where it apparently seeks dying bats for food. Raccoons are mostly solitary. Indeed, most procyonids are solitary, except

Striped hog-nosed skunk (*Conepatus semistriatus*)

the coatis. The coatis are seldom seen alone, but rather in groups of 15 or 20, often during the daylight hours in search of insects, sea turtle eggs, rodents, fruit, and a variety of other items.

The kinkajou is the most unusual of the family. Whereas all procyonids are good tree climbers, the kinkajou appears to be the most arboreal. In fact, many locals call it a "monkey of the night," and several common other names reflect this belief. The kinkajou alone among procyonids possesses a prehensile tail, which comes in handy for a species that spends its time in the branches.

Kinkajou (*Potos flavus*)

White-nosed coati (*Nasua narica*)

South American coati (*Nasua nasua*)

Crab-eating raccoon (*Procyon cancrivorus*)

11. Tapirs

Mammals of the order Perissodactyla are odd-toed hoofed animals. These are the hoofed animals that support most of their weight on their middle (third) toe, like horses, as opposed to cows (Artiodactyla), which appear to have a split hoof, but in reality are supporting their weight on two (third and fourth) toes. The Perissodactyla include the horses, rhinoceroses, and tapirs. Of the native wild mammals in South America only one genus belongs to this order, the genus *Tapirus*. Horses and donkeys have been reintroduced to South America by humans, and some wild populations exist in various regions.

The Tapiridae is the family of the tapirs, of which there are three species in South America. (Another species, *T. indicus,* is found in Southeast Asia.) Two of them are the largest of the land mammals in South America, weighing up to 250 kg (550 lbs). They are heavy bodied with a short, thick tail and short legs. The upper lip and nostrils are elongated, forming a mobile proboscis or trunk, but it is not nearly so long or useful as an elephant's trunk. The forefeet have four visible toes, but only three bear weight; the hind feet have only three toes.

Tapirs are semiaquatic and so are good swimmers; they even have the ability to swim underwater. Tapirs are primarily nocturnal, solitary animals and tend to avoid humans. They form evident trails through thick tropical vegetation for travel between their resting and feeding sites. Tapirs feed on herbaceous vegetation, grasses, and fruit. Like many ruminants, they will travel considerable distances to reach salt licks.

South American tapir (*Tapirus terrestris*)

12. Even-Toed Ungulates

The even-toed ungulates of the order Artiodactyla are the hoofed mammals that support their weight on the third and fourth toes. The result is a pair of hooves, rounded on the outer sides and straighter on the insides, sometimes referred to as cloven hooves. Usually there is a pair of smaller hooves behind the weight-supporting hooves, but these do not touch the ground. The even-toed ungulates include widely known species, such as deer, cattle, sheep, goats, pigs, giraffes, and hippopotamuses. In South America there are three families of even-toed ungulates; deer (family Cervidae), llamas (family Camelidae), and peccaries (family Tayassuidae).

All three of the South American families of artiodactyls eat plants, but peccaries are omnivorous and will also consume animal material when it is available. Both deer and llamas are ruminants with compartmentalized stomachs; thus, they "rechew" their food to aid in digestion. Peccaries have a less complicated, two-chambered stomach. Although the herbivorous lifestyle provides the advantage of an abundant food source (grass and other vegetation), that source has low nutrient value and the cell contents are protected by a cellulose cell wall. Because mammals have never evolved an ability to digest cellulose, herbivorous species, such as most artiodactyls, employ bacterial symbionts that have the ability to digest cellulose. The symbionts live in the gut and expose the plant cell contents, making them available for digestion and assimilation by the mammal. Herbivorous mammals also have specialized molars capable of grinding plant food and crushing and destroying the cellulose cell walls. Ruminant artiodactyls, deer and llamas, for example, graze food rapidly. The food passes to the first stomach (the rumen), where fermentation with symbiotic bacteria occurs. The deer and llamas then seek a resting site where they can regurgitate and "chew the cud." After chewing it extensively, it is swallowed and passed to other chambers of the stomach before passage into the intestine for assimilation.

Chewing all this plant matter wears the teeth. Thus, the degree of tooth wear has been a traditional method for estimating the age of animals such as deer. It is not very accurate, except in younger animals, however, as the amount of sand in the soil of some regions has an effect on the rate of tooth wear. As animals get very old, they may have teeth so worn that they have great trouble masticating their food.

Peccaries are unique to the New World, with all three species occurring in South America (two of the three species range into North America). Peccaries look quite a bit like pigs, but differ from them in several features. The first is the number and condition of the newborn. Pigs have large litters of young that are born with the eyes closed and are incapable of walking until they have grown and developed over several weeks of nursing and care. Peccaries give birth to fewer young, and the offspring are capable of walking and traveling with their mother immediately after birth. Another difference from pigs is that peccaries have tusks (upper canine teeth) that are directed downward, not laterally or upward as with pigs. Peccaries also have an extremely short tail. Unlike pigs, female peccaries have only one pair of mammary glands (pigs have many).

Peccaries possess a large, spongy scent gland on the rump, just in front of the tail. This gland emits a strong odor when the animal is excited. Curiously, the early European explorers believed that this gland was the navel, and as such it caused much commentary by early observers.

Collared peccary (*Pecari tajacu*)

White-lipped peccary (*Tayassu pecari*)

When peccaries live in arid conditions they grow extremely large kidneys. The kidneys also have greater medullary thickness and area. The change in kidney structure allows the peccaries to increase the concentration of electrolytes in their urine, thus allowing the animals to conserve water. In wetter environments, the urine is much more diluted and the peccaries can devote the energy that would have been devoted to kidney growth and wastewater processing into other areas (body growth, fat storage, reproduction, and other metabolic needs).

When one thinks of camels, the mind is immediately transported to Northern Africa (and parts of Asia, though that image is less crisp for most), yet South America boasts numerous members of the family Camelidae—the llamas, alpacas, vicuñas, and guanacos. Smaller than their Arabian relatives, South America's camelids live closer geographically to the ancestors of all camels. Camels arose in North America, perhaps 30 million years ago, and thereafter spread to Eurasia, Africa, and South America. Most scientists believe that camelids spread into other regions during times of low sea levels, probably during the Pliocene (5.3 to 1.8 million years ago).

In South America today there are two genera of Camelids. (There is a single genus in the Old World.) South American camelids do not have the humps seen in Old World camelids but do share the trait of being able to spit with considerable accuracy, even to a distance of 12 feet (4 m). The material ejected is evidently from the rumen, perhaps mixed with saliva, and leaves a burning sensation on the skin. The aim appears to be toward the eye, which can cause considerable irritation.

Unlike other mammals, the diaphragm contains a bone. This odd Y-shaped bone is small, only about 1.5 cm (1/2 inch), and is located just below the esophagus. Such a small bone, lacking articulation with any other bone, in such a large animal raises the question of whether it has a function or is the vestigial remains of a structure that had function in some ancient ancestor.

The two camelid species found in South America, guanacos and vicuñas, still have wild populations, but individual animals can be readily tamed. The familiar llamas and alpacas are found only as domestic animals and, in general, both are thought to be domestic versions of one or both of the two wild species. There is much debate on this point, with some arguing that there are actually four species, two wild and two—once wild—now completely domestic species. Some fossil evidence does support this view and the genetic evidence is difficult to interpret. Indeed, llamas can breed with guanacos and alpacas with vicuñas, with such coupling producing fertile offspring. The most recent taxonomic statements on this, including that of the respected Peter Grubb (in Wilson and Reeder, 2005), hold that llamas and alpacas are members of the single species, *Lama glama,* the guanacos.

The last group of artiodactyls is the deer (family Cervidae). Although South America does have large populations of introduced species of deer (red deer and elk) the thirteen species of native deer in South America are extraordinary. Excepting one species, all the native deer are endemic to

Llama (*Lama glama*)

Llama (*Lama glama*)

Guanaco
(*Lama glama*)
Joseph Gallen

Alpaca (*Lama glama*)

Vicuña (*Vicugna vicugna*)

the continent. Only the white-tailed deer (*Odocoileus virginianus*) is a species whose range expands outside the southern continent to North America.

There are six genera of native deer in South America, *Blastocerus* with one species, *Hippocamelus* with two species, *Mazama* with five species, *Odocoileus* (one species), *Ozotoceros* (one species), and *Pudu* with two species.

The marsh deer (*Blastocerus dichotomus*) is the largest of the South American deer, standing nearly 4 feet at the shoulders and weighing up to 275 lbs (125 kg). The antlers of the adult male are doubly forked, resulting in a total of four points on each antler. The two toes, bound by a membrane, can be spread up to 10 cm, more than any other South American deer, an adaptation to getting around in swamps. The color above in summer is a bright rufus chestnut turning in winter to brownish red, below the color is somewhat lighter, and the lower parts of the legs are black. They prefer swamps where they seek out wet islands on which to rest. They are very timid, found in groups of 2 to 5, occasionally even grazing with domestic livestock, in the evenings and during the night. Marsh deer are quite rare in present times, in part, because hunters prize them for their meat and their susceptibility to cattle diseases.

Marsh deer (*Blastocerus dichotomus*)

Peruvian huemul (*Hippocamelus antisensis*)
Mammal Image Library

A medium deer, the Peruvian huemul (*Hippocamelus antisensis*) reaches about 30 inches (87 cm) at the shoulders. It is grayish bay, sprinkled with dark hairs, and its black face distinguishes it from other deer in its range. The male's horns are each forked with two long points. They are found on the rocky steppes between 3,000 and 5,000 m altitude in the Andes of Peru, northern Chile, and northwestern Argentina. Unlike other South American deer species, the Peruvian huemul does not inhabit forests.

The Patagonian huemul (*Hippocamelus bisulcus*) is a deer of the southern Andes of Chile and Argentina. It is relatively robust, reaching about 30 inches (79 cm) in height at the shoulders and weighing about 150 lbs (70 kg). The male horns never develop beyond a single fork. The Patagonian huemuls are dark brown, somewhat tan in summer and grayish in winter. They associate in groups of 2 to 8, living at higher elevations in the summer, at or near the tree line, and descending to the valleys in the winter. Preferring steep, rocky slopes with dense shrubs and forest clearings, they are primarily diurnal but may be active at night. This species is in danger of extinction because of intensive hunting, predation by dogs, susceptibility to cattle diseases, and competition with cattle and the introduced red deer.

Red brockets (*Mazama americana*) are small deer with ears shorter than in white-tailed deer. In males, the antlers are unbranched, simple spikes. Above the color is a deep reddish brown, the anterior part of the lips is white, the ventral surface of the body is rusty to reddish brown, sprinkled with white hairs. The chin and upper throat are whitish (but not pure white as in *white-tailed deer*). Red brockets reach a height of about 30 inches (76 cm) at the shoulders and may weigh up to 100 lbs (48 kg). They prefer the dense tropical forest, are solitary or found in pairs, and are largely crepuscular and nocturnal. They appear to breed throughout the year in the tropics, frequently giving birth to twins. The fawns are born with white spots that they will lose in about 2–3 months. Although they are common, they are seldom seen because of their secretive habits. Though they may be active by day, they are primarily seen by night, and usually as solitary individuals.

Red brocket deer
(*Mazama americana*)

Gray brocket deer
(*Mazama gouazoubira*)

The gray brocket (*Mazama gouazoubira*) is a small deer with ears shorter than in the white-tailed deer, and in males, the antlers are unbranched, simple spikes. They are a dull grayish brown above (not reddish as in the *red brocket*). This species is somewhat smaller than the red brocket, reaching about 24 inches (61 cm) at the shoulders and weighing about 40 lbs (18 kg). It is mainly diurnal and crepuscular and is solitary or found in pairs. They are found in dense humid forest from Colombia and Venezuela south through Brazil to northern Argentina. Seldom seen, they seek out the dense streamside vegetation in rain forests.

The well-known white-tailed deer (*Odocoileus virginianus*) is a large deer with large ears and, in the male, antlers that consist of a curved main beam with several prongs issuing from it. These prongs never branch as they do in the marsh deer. In Brockets and Pudu the antlers consist of unbranched simple spikes. The color above is tawny ochreous brown, tending slightly to gray, posteriorly the color becomes more rufus. The ventral surface shows a pure white area between the fore and hind legs. White-tailed deer are next in size to the marsh deer, reaching about 30 inches (80 cm) at the shoulders and weighing 90 to 100 lbs (40 to 50 kg), making

White-tailed deer (*Odocoileus virginianus*)

them much smaller than their species achieves in North America. They are diurnal, nocturnal, and crepuscular depending on region, preferring the forest edge but found in open llanos, feeding in clearings and llanos, social, often found in family groups.

The most elegant and gracious of South American deer is the pampas deer (*Ozotoceros bezoarticus*); its fur is short and smooth as satin. It is bay or reddish bay with tendencies toward gray in some regions, and below it is white. Males have hardened antlers from November to July. Usually one-year-old males have single pointed antlers, two-year-olds have a total of four points, and three-year-olds have six points total. They are medium, reaching about 38 inches (95 cm) at the shoulders and weighing about 75 lbs (34 kg). They can be found not only in flat land and open fields, avoiding trees and brushland, but also in rolling hills and temporarily inundated areas. They are active during the day and night and can be found in small groups of 5 to 6, but they are most often seen as solitary animals. In Brazil the young are born from July to December, while in Argentina reproduction is from December to February. Their numbers are diminishing because of hunting and their susceptibility to foot-and-mouth disease.

Pampas deer (*Ozotoceros bezoarticus*)
Mammal Image Library

Southern pudu
(*Pudu puda*)
Mammal Image Library

The southern pudu (*Pudu puda*), a very small deer, stands only 15 inches (38 cm) at its shoulders and weighs only about 22 lbs (10 kg). In the males the horns are very short and simple. The color varies from rufus to dark brown or gray. Shy and fleet, they inhabit the dense forest of the Andean foothills of southern Chile and southwestern Argentina. They are active in both the morning and evening, they are solitary or found in small groups of two or three animals, feeding on twigs, buds, flowers, grass, and other vegetation. They are considered to be the smallest deer in the world. Because of their shy habits they are seldom seen in nature, but although scarce, they are not considered to be in danger of extinction.

Male deer are noted for their deciduous antlers, which they shed annually. Female deer do not grow antlers and are usually smaller than the males. The male's antlers first grow out when the deer is about two years of age. Although some species have but one prong, the fully adult males of many species possess multiple prongs. Young males may have only one prong in their second year, and in subsequent years grow more, but the number of prongs is not a measure of the age of the deer. Deer receiving adequate nutrition will grow well-formed heavy antlers, and those eating poorly may have thin spindly antlers. (Their food varies from browsing of shoots and buds of bushes to grazing on grasses.) When new antlers are growing they are soft, tender, and covered with a fuzzy skin, called the velvet. When the antlers are large and well formed, the blood circulation to the velvet is cut off and it dries, causing an evident irritation or itching, resulting in the deer then rubbing off this dried skin by thrashing their antlers against bushes and small trees. Males use their antlers in battles with others of their own species to determine dominance and thus which male retains a harem with which to copulate and produce offspring. Antlers can also be used as defensive weapons against predators.

Deer are born to run. The strong hooves on their third and fourth toes give them an advantage. These hooves are kept sharp through use and as such they slightly penetrate the ground when they are running. Only rarely do they slip. Young deer are precocial, they are up on their feet in less than an hour after birth and soon thereafter are following their mothers. Young deer are often spotted and their coloration provides better camouflage. Very young deer are frequently left by their mothers to rest, confidant that they will remain undetected by predators. It appears that resting deer do not liberate odor and it is difficult for predators to locate them by smell. But moving deer do leave a scent trail and are readily followed by persistent predators. The activity patterns of the white-tailed deer in South America are opposite to those of the white-tailed deer in North America. In North America white-tailed deer are basically nocturnal, even though they are often seen in the early morning or evening. On the llanos of Venezuela, however, their most active period is at high noon. There, they are seldom seen at night.

13. Whales and Dolphins

Whales, dolphins, and porpoises are all members of the order Cetacea, thus the common name for this group is the cetaceans or, inappropriately, marine mammals. The reasons that marine mammals and cetaceans are not interchangeable are twofold. First, the marine mammals also include sea lions, seals, otters, manatees and other species. Second, not all cetaceans are marine, most notably the river dolphins of South America.

Of all mammals, the cetaceans are the best adapted for life in the water. Their smooth, virtually hairless bodies and fishlike shape allow them to swim with ease. Equally important are the location of their breathing passageway, or blow hole, on the dorsal side of the body and their ability to remain submerged for minutes at a time and easily inhale after surfacing. Finally, the body of many cetaceans has many morphological and physiological adaptations: the animals can withstand the pressure of deep dives; their brains can withstand anoxia; the forelimbs are modified to act as oars; the hind limbs are absent; and the tail is flattened horizontally, broadened and moved vertically to provide propulsion.

The inland waters of South America contain two of the world's five river dolphins. These dolphins are the gray dolphin (family Delphinidae), which also is found in coastal waters, and the pink river dolphin (family Iniidae), which is more restricted to rivers. Some studies suggest they adapted to river life about 15 million years ago. The Iniidae look quite different from the more familiar dolphins, with a blunt forehead having a pronounced bulge, a long slender beak, and a long, low dorsal fin. The flexible neck is quite distinct, having unfused vertebrae. Pink river dolphins are seen singly, in pairs or groups of up to 6 or more, and are active by both day and night. During periods of high water they invade small streams and inundated lands, sometimes becoming fatally trapped when waters recede.

Pink river dolphin (*Inia geoffrensis*)

The gray dolphins possess a snout that is beaklike and not well defined. The dorsal fin is triangular and prominent. In short, they appear much more like the typical "dolphin" most of us recognize.

Whales are placental mammals that live their entire lives in the seas; they eat, sleep, copulate, and give birth to their young without leaving the water. Whales are conveniently subdivided into two suborders, the toothed whales (suborder Odontoceti) and the baleen whales (suborder Mysticeti). The toothed whales possess teeth rather than baleen and they pursue their prey utilizing echolocation. In the world there are 71 species and 7 families of toothed whales. Of these families, the beaked whales of the family Ziphiidae with 21 species worldwide, has 6 species living in the waters surrounding South America. Beaked whales have sometimes been involved in mass strandings. Groups of whales come ashore for unknown reasons and usually die. Speculations that have been advanced vary from suicide to damage to the inner ear caused by either military sonar devices or parasitic infections. Also the sperm whales (family Physeteridae), with three species worldwide, has only one species living around South America. Sperm whales feed primarily on squid, but also consume octopus and fish. They have always been keenly sought after by whalers, and their numbers are dwindling, despite protection. They are listed as vulnerable. Killer whales, false killer whales, and oceanic dolphins belong to the family Delphinidae, which has 34 species worldwide, but only 9 species residing in South American waters. Both killer whales (also called orca) and false killer whales are favorite species for training for aquarium shows.

Baleen whales (suborder Mysticeti) possess baleen instead of teeth. Baleen is a hornlike material that is formed like the teeth of a comb, with its inner surface forming fine hairlike material that serves to strain food from the water. There are four families, the rorquals, right whales, pigmy right whales, and gray whales (none of the latter found near South America). The rorquals of the family Balaenopteridae have a total of seven species worldwide, and five of these are found in South American waters. They are fast swimming, streamlined whales with visible grooves on their throats as well as dorsal fins. They feed on fish. On the other hand, the right whales (family Balaenidae) are slow swimming, feeding by skimming, primarily on small invertebrates and some small fish. There are four species worldwide, but only one species near South America, the famous southern right whale. This species has been protected in Argentine waters and its numbers are clearly recovering. Right whales put on a show, breaching (jumping out of the water) and lobtailing, or smacking the surface of the water with its tail. Finally the pigmy right whale belongs to the family Neobalaenidae which has only one species worldwide and it resides in South American waters. This species is scarce and rarely observed. They are slow swimmers, feeding on very small crustaceans (copepods).

Killer whale (*Orcinus orca*)

Guanacos

SPECIES DISTRIBUTION TABLE

Each species of South American mammals is included in the table that begins on the following page. Within the class Mammalia, the first taxonomic subdivision is the order, followed by the family. Orders are arranged according to the chapters in this book, which in turn follows the order presented by Wilson and Reeder, *Mammal Species of the World,* third edition. When possible, a common name is provided for orders, families, and in some instances subcategories of families. Thereafter the genus and species are shown, listed in alphabetical order. Whales and oceanic dolphins are not listed in the table as they do not occur within individual countries.

* species present P possible i introduced x extirpated	Argentina	Brazil	Bolivia	Chile	Colombia	Ecuador	French Guiana	Guyana	Paraguay	Peru	Suriname	Trinidad	Uruguay	Venezuela
Order Didelphimorphia (marsupials, in part)														
Family Didelphidae (Opossums)														
Woolly opossoms														
Caluromys derbianus					*	*								
Caluromys lanatus	*	*	*		*	*			*	*				*
Caluromys philander		*	*				*	*			*	*		*
Caluromysiops irrupta		*			*					*				
Glironia venusta		*	*			*				*				
Opossums														
Chironectes minimus	*	*			*	*	*	*	*	*	*			*
Didelphis albiventris	*	*	*						*				*	
Didelphis aurita	*	*							*					*
Didelphis imperfecta		*					*				*			*
Didelphis marsupalis		*	*		*	*	*	*		*	*	*		*
Didelphis pernigra			*		*	*				*				*
Lutreolina crassicaudata	*	*	*		*		*	*	*		*		*	*
Philander andersoni					*	*				*				*
Philander frenatus	*	*							*					
Philander mcilhennyi		*								*				
Philander opossum		*			*	*	*	*	*	*	*	*		*
Brown four-eyed opossum														
Metachirus nudicaudatus	*	*	*		*	*	*	*	*	*	*			*
Gracile opossums, mouse opossums, and short-tailed opossums														
Gracilinanus aceramarcae			*							*				
Gracilinanus agilis	*	*	*						*	*			*	
Gracilinanus agricolai		*												
Gracilinanus dryas														*

* species present P possible i introduced x extirpated	Argentina	Brazil	Bolivia	Chile	Colombia	Ecuador	French Guiana	Guyana	Paraguay	Peru	Suriname	Trinidad	Uruguay	Venezuela
Gracilinanus emiliae		*			*		*	*			*			*
Gracilinanus formosus	*													
Gracilinanus ignitus	*													
Gracilinanus marica					*									*
Gracilinanus microtarsus		*												
Hyladelphys kalinowski							*	*		*				
Lestodelphys halli	*													
Marmosa andersoni										*				
Marmosa lepida		*	*		*	*				*	*			
Marmosa murina		*			*	*	*	*		*	*	*		*
Marmosa quichua										*				
Marmosa robinsoni					*	*				*				*
Marmosa rubra						*				*				
Marmosa tyleriana														*
Marmosa xerophila					*									*
Marmosops bishopi		*	*							*				
Marmosops cracens														*
Marmosops dorothea			*											
Marmosops fuscatus					*							*		*
Marmosops handleyi					*									
Marmosops impavidus		*	*		*	*				*				*
Marmosops incanus		*												
Marmosops juninensis										*				
Marmosops neblina		*				*				*				*
Marmosops noctivagus		*	*			*				*				
Marmosops parvidens		*			*		*	*						*
Marmosops paulensis		*												
Marmosops pinheiroi		*					*	*			*			*
Micoureus alstoni				P										
Micoureus constantiae	*	*	*											

	Argentina	Brazil	Bolivia	Chile	Colombia	Ecuador	French Guiana	Guyana	Paraguay	Peru	Suriname	Trinidad	Uruguay	Venezuela
* species present P possible i introduced x extirpated														
Micoureus demerarae		*	*		*	*	*	*		*	*			*
Micoureus paraguayanas		*							*					
Micoureus phaeus					*	*								
Micoureus regina		*	*		*	*				*				
Monodelphis adusta			*		*	*				*				
Monodelphis americana		*												
Monodelphis brevicaudata		*					*	*			*			*
Monodelphis dimidiata	*	*											*	
Monodelphis domestica	*	*	*						*					
Monodelphis emiliae		*	*							*				
Monodelphis glirina		*	*							*				
Monodelphis iheringi		*												
Monodelphis kunsi		*	*											
Monodelphis maraxina		*												
Monodelphis osgoodi			*							*				
Monodelphis palliolata					*									*
Monodelphis rubida		*												
Monodelphis scalops		*							*					
Monodelphis sorex	*	*							*					
Monodelphis theresa		*												
Monodelphis umbistriata		*												
Monodelphis unistriata	*	*												
Thylamys cinderella	*		P											
Thylamys elegans				*										
Thylamys karimii		*												
Thylamys macrurus		*							*					
Thylamys pallidior	*		*	*						*				
Thylamys pusillus	*		*						*					
Thylamys sponsorius	*		*											
Thylamys tatei										*				

	Argentina	Brazil	Bolivia	Chile	Colombia	Ecuador	French Guiana	Guyana	Paraguay	Peru	Suriname	Trinidad	Uruguay	Venezuela
* species present P possible i introduced x extirpated														
Thylamys vetulinus		*												
Thylamys venustus	*		*											
Order Paucituberculata (marsupials, in part)														
Family Caenolestidae (Shrew opossums)														
Caenolestes caniventer						*				*				
Caenolestes condorensis						*								
Caenolestes convelatus					*	*								
Caenolestes fuliginosus					*	*								*
Lestoros inca			*							*				
Rhyncholestes raphanurus	*			*										
Order Microbiotheria (Monito del Monte)														
Family Microbiotheriidae														
Dromiciops gliroides	*		*											
Order Sirenia														
Family Trichechidae (Manatees)														
Trichechus inunguis		*			*	*		*		*				
Trichechus manatus		*			*		*	*			*	*		*
Order Cingulata														
Family Dasypodidae (Armadillos)														
Long-nosed armadillos														
Dasypus hybridus	*	*							*				*	
Dasypus kappleri		*	*		*	*	*	*		*	*			*
Dasypus novemcinctus	*	*	*		*		*	*	*	*	*	*		*
Dasypus pilosus										*				
Dasypus sabanicola					*									*
Dasypus septemcinctus	*	*	*						*					
Dasypus yepesi	*													

* species present P possible i introduced x extirpated	Argentina	Brazil	Bolivia	Chile	Colombia	Ecuador	French Guiana	Guyana	Paraguay	Peru	Suriname	Trinidad	Uruguay	Venezuela
Fairy armadillos														
Calyptophractus retusus	*		*						*					
Chlamyphorus truncatus	*													
Hairy armadillos														
Chaetophractus nationi			*	*										
Chaetophractus vellerosus	*		*	*					*					
Chaetophractus villosus	*		*	*					*					
Euphractus sexcinctus	*	*	*						*		*		*	
Rhyncholestes raphanurus	*			*										
Zaedyus pichiy	*			*										
Naked-tailed armadillos and giant armadillo														
Cabassous centralis					*									*
Cabassous chacoensis	*								*					
Cabassous tatouay	*	*											*	
Cabassous unicinctus		*	*		*	*				*				
Priodontes maximus	*	*	*		*	*	*	*	*	*	*			*
Three-banded armadillos														
Tolypeutes matacus	*	*	*						*					
Tolypeutes tricinctcus		*												
Order Pilosa (Sloths and anteaters)														
Family Bradypodidae (Three-toed sloths)														
Bradypus torquatus		*												
Bradypus tridactylus		*					*	*			*			*
Bradypus variegatus	*	*	*		*	*	*	*	*	*	*	*		*
Family Megalonichidae (Two-toed sloths)														
Choloepus didactylus		*			*	*	*	*		*	*			*
Choloepus hoffmanni		*	*		*	*								*

* species present P possible i introduced x extirpated	Argentina	Brazil	Bolivia	Chile	Colombia	Ecuador	French Guiana	Guyana	Paraguay	Peru	Suriname	Trinidad	Uruguay	Venezuela
Family Cyclopedidae (Silky anteater)														
Cyclopes didactylus		*	*		*	*	*	*		*	*	*		*
Family Myrmecophagidae (Anteaters)														
Myrmecophaga tridactyla	*	*	*		*	*	*	*	*	*		*		*
Tamandua mexicana					*	*				*				*
Tamandua tetradactyla	*	*	*		*	*	*	*	*	*	*		*	*
Order Primates (Monkeys, apes, etc.)														
Family Cebidae (Monkeys)														
Marmosets and tamarins														
Callimico goeldii		*	*		*	*				*				
Callithrix acariensis		*												
Callithrix argentata		*	*											
Callithrix aurita		*												
Callithrix chrysoleuca		*												
Callithrix emiliae		*												
Callithrix flaviceps		*												
Callithrix geoffroyi		*												
Callithrix humeralifera		*												
Callithrix humilis		*												
Callithrix intermedia		*												
Callithrix jacchus		*												
Callithrix kuhlii		*												
Callithrix leucippe		*												
Callithrix manicorensis		*												
Callithrix marcai		*												
Callithrix mauesi		*												
Callithrix melanura		*	*											

* species present P possible i introduced x extirpated	Argentina	Brazil	Bolivia	Chile	Colombia	Ecuador	French Guiana	Guyana	Paraguay	Peru	Suriname	Trinidad	Uruguay	Venezuela
Callithrix nigriceps		*												
Callithrix pennicillata		*												
Callithrix pygmaea		*				*				*				
Callithrix saterei		*												
Leontopithecus caissara		*												
Leontopithecus chrysomelas		*												
Leontopithecus chrysopygus		*												
Leontopithecus rosalia		*												
Saguinus bicolor		*								P				
Saguinus fuscicollis		*	*		*	*				*				
Saguinus geoffroyi					*									
Saguinus graellsi					*	*				*				
Saguinus imperator		*	*							*				
Saguinus inustus		*			*									
Saguinus labiatus		*	*							*				
Saguinus leucopus					*					*				*
Saguinus martinsi		*												
Saguinus melanoleucus		*												
Saguinus midas		*					*	*			*			
Saguinus mystax		*								*				
Saguinus niger		*												
Saguinus nigricollis		*				*				*				
Saguinus oedipus					*									
Saguinus pileatus		*												
Saguinus tripartitus		*				*								
Capuchin monkeys														
Cebus albifrons		*	*		*	*				*		*		*
Cebus apella	*	*			*	*	*	*	*	*	*	*		*
Cebus capucinus					*	*								
Cebus kaapori		*												

* species present P possible i introduced x extirpated	Argentina	Brazil	Bolivia	Chile	Colombia	Ecuador	French Guiana	Guyana	Paraguay	Peru	Suriname	Trinidad	Uruguay	Venezuela
Cebus libidinosus		*	*						*					
Cebus nigritus		*												
Cebus olivaceus		*			P		*	*				*	*		*
Cebus xanthosternos		*												
Squirrel monkeys														
Saimiri boliviensis		*	*							*				
Saimiri sciureus		*			*	*	*	*		*	*			*
Saimiri ustus		*												
Saimiri vanzolinii		*												
Family Aotidae (Night monkeys)														
Aotus azarae	*		*						*					
Aotus hershkovitzi					*									
Aotus lemurinus					*	*								
Aotus miconax										*				
Aotus nancymaae		*								*				
Aotus nigriceps		*								*				
Aotus trivirgatus		*												*
Aotus vociferans		*			*									
Family Pitheciidae														
Titi monkeys														
Callicebus baptista		*												
Callicebus barbarabrownae		*												
Callicebus bernhardi		*												
Callicebus brunneus		*	*							*				
Callicebus caligatus		*												
Callicebus cinerascens		*												
Callicebus coimbrai		*												
Callicebus cupreus		*	P							*				
Celicebus discolor		P			*	*				*				

* species present P possible i introduced x extirpated	Argentina	Brazil	Bolivia	Chile	Colombia	Ecuador	French Guiana	Guyana	Paraguay	Peru	Suriname	Trinidad	Uruguay	Venezuela
Callicebus donacophilus			*											
Callicebus dubius		*												
Callicebus hoffmannsi		*												
Callicebus lucifer		*				*				*				
Callicebus lugens		*			*									*
Callicebus medemi					*									
Callicebus melanochir		*												
Callicebus modestus			*											
Callicebus moloch		*												
Callicebus nigrifrons		*												
Callicebus oenanthe										*				
Callicebus olallae			*											
Callicebus ornatus					*									
Callicebus pallescens		*	P						*					
Callicebus personatus		*												
Callicebus purinus		*												
Callicebus regulus		*												
Callicebus stephennashi		*												
Callicebus torquatus		*												
Saki and uakari monkeys														
Cacajao calvus		*								*				
Cacajao melanocephalus		*												*
Chiropotes albinasus		*												
Chiropotes chiropotes		*					*	*			*			
Chiropotes israelita		*												*
Chiropotes satanus		*												
Chiropotes utahickae		*												
Pithecia aequatorialis						*				*				
Pithecia albicans		*												

* species present P possible i introduced x extirpated	Argentina	Brazil	Bolivia	Chile	Colombia	Ecuador	French Guiana	Guyana	Paraguay	Peru	Suriname	Trinidad	Uruguay	Venezuela
Pithecia irrorata		*	*							*				
Pithecia monachus		*			*	*				*				
Pithecia pithecia		*					*	*			*			*
Family Atelidae														
Howler monkeys														
Alouatta belzebul		*												
Alouatta caraya	*	*	*											
Alouatta guariba		*	*											
Alouatta macconnelli		*					*	*			*	*		
Alouatta nigerrima		*												
Alouatta palliata					*	*								
Alouatta sara			*											
Alouatta seniculus		*			*									*
Spider monkeys														
Ateles belzebuth					*					*				*
Ateles chamek		*	*							*				
Ateles fuscipes					*	*								
Ateles geoffroyi					P									
Ateles hybridus					*									*
Ateles marginatus		*												
Ateles paniscus		*					*	*			*			
Brachyteles arachnoides		*												
Lagothrix cana		*	*							*				
Lagothrix lagotricha		*			*	*				*				
Lagothrix lugens					*									*
Lagothrix poeppigii		*				*				*				
Oroenax flavicauda		*								*				
Order Rodentia (Rodents)														
Family Sciuridae (Squirrels)														

* species present P possible i introduced x extirpated	Argentina	Brazil	Bolivia	Chile	Colombia	Ecuador	French Guiana	Guyana	Paraguay	Peru	Suriname	Trinidad	Uruguay	Venezuela
Microsciurus alfari					*									
Microsciurus flaviventer		*			*	*				*				
Microsciurus mimulus					*	*								
Microsciurus santanderensis					*									
Sciurillus pusillus		*					*			*	*			
Sciurillus aestuans		*					*	*			*			*
Sciurillus flammifer														*
Sciurillus gilvigularis		*						*						*
Sciurillus granatensis					*	*						*		*
Sciurillus ignitus	*	*	*							*				
Sciurillus igniventris		*			*	*				*				*
Sciurillus pucheranii					*									
Sciurillus pyrrhinus										*				
Sciurillus sanborni										*				
Sciurillus spadiceus		*	*		*	*				*				
Sciurillus stramineus						*				*				
Family Heteromyidae (Spiny pocket mice)														
Heteromys anomalous					*							*		*
Heteromys australis					*	*								
Heteromys desmarestianus					*									
Heteromys oascicus														*
Heteromys teleus						*								
Family Muridae (Old World rats and mice)														
Mus musculus	i	i	i	i	i	i	i	i	i	i	i	i	i	i
Rattus norvegicus	i	i	i	i	i	i	i	i	i	i	i	i	i	i
Rattus rattus	i	i	i	i	i	i	i	i	i	i	i	i	i	i
Family Cricetidae														
Harvest mice														

* species present P possible i introduced x extirpated	Argentina	Brazil	Bolivia	Chile	Colombia	Ecuador	French Guiana	Guyana	Paraguay	Peru	Suriname	Trinidad	Uruguay	Venezuela
Isthmomys pirrensis					P									
Reithrodontomys darienensis					P									
Reithrodontomys mexicanus					*	*								
Rats and mice														
Abrawayaomys ruschii	*	*												
Abrothrix andinus	*		*	*						*				
Abrothrix hershkovitzi				*										
Abrothrix illuteus	*													
Abrothrix jelskii	*		*							*				
Abrothrix lanosus	*			*										
Abrothrix longipilis	*			*										
Abrothrix markhami				*										
Abrothrix olivaceus	*			*										
Abrothrix sanborni	P			*										
Aepeomys lugens					*	*								*
Aepeomys reigi														*
Akodon aerosus			*			*				*				
Akodon affinis					*									
Akodon albiventer	*		*	*						*				
Akodon aliquantulus	*													
Akodon azarae	*	*							*				*	
Akodon bogotensis					*									*
Akodon boliviensis			*							*				
Akodon budini	*		*											
Akodon cursor	*	*												
Akodon dayi			*											
Akodon dolores	*													
Akodon fumeus	*		*							*				
Akodon iniscatus	*													

	Argentina	Brazil	Bolivia	Chile	Colombia	Ecuador	French Guiana	Guyana	Paraguay	Peru	Suriname	Trinidad	Uruguay	Venezuela
* species present P possible i introduced x extirpated														
Akodon juninensis										*				
Akodon kofordi			*							*				
Akodon latebricola						*								
Akodon leocolimnaeus	*													
Akodon lindberghi		*												
Akodon lutescens	*		*							*				
Akodon mimus			*							*				
Akodon molinae	*													
Akodon mollis						*				*				
Akodon montensis	*	*							*					
Akodon mystax		*												
Akodon neocenus	*													
Akodon oenos	*													
Akodon orophilus										*				
Akodon paranaensis	*	*												
Akodon pervalens	P		*											
Akodon philipmyersi	*													
Akodon reigi		*											*	
Akodon sanctipaulensis		*												
Akodon serrensis		*												
Akodon siberiae			*											
Akodon simulator	*		*											
Akodon spegazzinii	*													
Akodon subfuscus			*							*				
Akodon surdus										*				
Akodon sylvanus	*													
Akodon toba	*		*						*					
Akodon torques										*				
Akodon varius			*											

* species present P possible i introduced x extirpated	Argentina	Brazil	Bolivia	Chile	Colombia	Ecuador	French Guiana	Guyana	Paraguay	Peru	Suriname	Trinidad	Uruguay	Venezuela
Amphinectomys savamis										*				
Andalgalomys olrogi	*													
Andalgalomys pearsoni			*						*					
Andalgalomys roigi	*													
Andinomys edax	*		*	*						*				
Anotomys leander						*								
Auliscomys boliviensis			*	*						*				
Auliscomys pictus			*							*				
Auliscomys sublimus	*		*	*						*				
Bibimys chacoensis	*													
Bibimys labiosus		x												
Bibimys torresi	*													
Blarinomys breviceps	*	*												
Brucepattersonius albinasus		*												
Brucepattersonius griserufescens		*												
Brucepattersonius guarani	*													
Brucepattersonius igniventris		*												
Brucepattersonius iheringi	*	*												
Brucepattersonius misionensis	*													
Brucepattersonius paradisus	*													
Brucepattersonius soricinus		*												
Calomys boliviae	*		*											
Calomys callidus	*								*					
Calomys callosus	*	*	*						*					
Calomys expulsus		*												
Calomys hummelincki					*									*
Calomys laucha	*	*	*						*				*	
Calomys lepidus	*		*	*						*				
Calomys musculinus	*		*						*					

* species present P possible i introduced x extirpated	Argentina	Brazil	Bolivia	Chile	Colombia	Ecuador	French Guiana	Guyana	Paraguay	Peru	Suriname	Trinidad	Uruguay	Venezuela
Calomys sorellus										*				
Calomys tener	*	*	*											
Calomys tocantinsi		*												
Calomys venustus	*													
Chelemys delfini				*										
Chelemys macronyx	*			*										
Chelemys megalonyx				*										
Chibchanomys orcesi						*								
Chibchanomys trichotis					*					*				*
Chilomys instans				*	*									*
Chinchillula sahamae			*	*						*				
Delomys collinus		*												
Delomys dorsalis	*	*												
Delomys sublineatus		*												
Deltamys kempi	*	*											*	
Eligmodontia moreni	*													
Eligmodontia morgani	*			*										
Eligmodontia puerulus	*		*	*						*				
Eligmodontia typus	*													
Euneomys chinchilloides				*										
Euneomys fossor	*													
Euneomys mordax	*			*										
Euneomys petersoni	*			*										
Galenomys garleppi			*	*						*				
Geoxus valdivianus	*			*										
Graomys centralis	*													
Graomys domorum	*		*											
Graomys edithae	*													
Graomys griseoflavus	*	*	*						*					

	Argentina	Brazil	Bolivia	Chile	Colombia	Ecuador	French Guiana	Guyana	Paraguay	Peru	Suriname	Trinidad	Uruguay	Venezuela
* species present P possible i introduced x extirpated														
Handleyomys fuscatus					*									
Handleyomys intectus					*									
Holochilus brasiliensis	*	*											*	
Holochilus chacarius	*								*					
Holochilus sciureus		*	*		*	*	*	*		*	*			*
Ichthyomys hydrobates					*	*								*
Ichthyomys pittieri														*
Ichthyomys stolzmanni						*				*				
Ichthyomys tweedii					P	*								
Irenomys tarsalis	*			*										
Juliomys pictipes	*	*												
Juliomys rimofrons		*												
Juscelinomys candango		*												
Juscelinomys guaporensis			*											
Juscelinomys huanchacae			*											
Kunsia fronto	*	*												
Kunsia tomentosus		*	*											
Lenoxus apicalis			*							*				
Loxodontomys micropus	*			*										
Loxodontomys pikumche				*										
Lundomys molitor		*											*	
Melanomys caliginosus					*	*								*
Melanomys robustulus						*								
Melanomys zunigae										*				
Microakodontomys transitorius		*												
Microryzomys altissimus					*	*				*				
Microryzomys minutus			*		*	*				*				*
Neacomys dubosti		*					*				*			
Neacomys guianae		*					*	*			*			*

	Argentina	Brazil	Bolivia	Chile	Colombia	Ecuador	French Guiana	Guyana	Paraguay	Peru	Suriname	Trinidad	Uruguay	Venezuela
* species present P possible i introduced x extirpated														
Neacomys minutus		*												
Neacomys musseri		*								*				
Neacomys paracou		*					*	*			*			*
Neacomys spinosus		*	*		*	*				*				
Neacomys tenuipes					*									*
Necromys amoenus			*							*				
Necromys benefactus	*													
Necromys lactens	*		*											
Necromys lasiurus	P	*							*					
Necromys lenguarum	P		*						*					
Necromys obscurus	*												*	
Necromys punctulatus					P	*								
Necromys temchuki	*													
Necromys urichi		*			*							*		*
Nectomys apicalis		*	*			*				*				
Nectomys magdalenae					*									
Nectomys palmipes												*		*
Nectomys rattus		*			*		*	*		P	*			*
Nectomys squamipes	*	*							*					
Nectomys ebriosus	*		*	*						*				
Neusticomys ferreirai		*												
Neusticomys monticolus					*	*								
Neusticomys mussoi														*
Neusticomys oyapocki		*					*							
Neusticomys peruviensis										*				
Neusticomys venezuelae								*						*
Notiomys edwardsii	*													
Oecomys auyantepui		*					*	*			*			*
Oecomys bicolor		*	*		*	*	*	*		*	*			*

	Argentina	Brazil	Bolivia	Chile	Colombia	Ecuador	French Guiana	Guyana	Paraguay	Peru	Suriname	Trinidad	Uruguay	Venezuela
* species present P possible i introduced x extirpated														
Oecomys catherinae		*												
Oecomys cleberi		*												
Oecomys concolor		*	*		*									*
Oecomys flavicans					*									*
Oecomys mamorae		*	*						*					
Oecomys paricola		*												
Oecomys phaeotis										*				
Oecomys rex		*					*	*			*			*
Oecomys roberti		*	*				*	*		*	*			*
Oecomys rutilus		*					*	*			*			*
Oecomys speciosus					*							*		*
Oecomys superans		P			*	*				*				
Oecomys trinitatus		*			*	*	*	*		*	*	*		
Oligoryzomys andinus			*							*				
Oligoryzomys arenalis										*				
Oligoryzomys brendae	*													
Oligoryzomys chacoensis	*	*	*						*					
Oligoryzomys delticola	*	*											*	
Oligoryzomys destructor	*		*		*	*				*				
Oligoryzomys eliurus		*												
Oligoryzomys flavescens	*	*							*				*	
Oligoryzomys fornesi	*	*							*					
Oligoryzomys fulvescens		*			*	*	*	*			*			*
Oligoryzomys griseolus					*									*
Oligoryzomys longicaudatus	*			*										
Oligoryzomys magellanicus	*			*										
Oligoryzomys microtis		*	*						*	*				
Oligoryzomys nigripes	*	*							*					
Oligoryzomys stramineus		*												

	Argentina	Brazil	Bolivia	Chile	Colombia	Ecuador	French Guiana	Guyana	Paraguay	Peru	Suriname	Trinidad	Uruguay	Venezuela
* species present P possible i introduced x extirpated														
Oryzomys albigularis					*	*				*				*
Oryzomys alfaroi					*	*								
Oryzomys angouya	*	*							*					
Oryzomys auriventer						*				*				
Oryzomys balneator						*				*				
Oryzomys bolivaris					*	*								
Oryzomys caracolus														*
Oryzomys couesi					*									
Oryzomys curasoae														P
Oryzomys emmonsae		*												
Oryzomys gorgasi					*									*
Oryzomys hammondi						*								
Oryzomys keaysi										*				
Oryzomys lamia		*												
Oryzomys laticeps		*												
Oryzomys legatus	*		*											
Oryzomys levipes			*							*				
Oryzomys macconnelli		*			*	*	*	*		*	*			*
Oryzomys maracajuensis		*							P					
Oryzomys marinhus		*												
Oryzomys megacephalus		*					*	*	*		*	*		*
Oryzomys meridensis														*
Oryzomys nitidus		*	*							*				
Oryzomys perenensis		*	*		*	*				*				
Oryzomys russatus	*	*							*					
Oryzomys scotti		*												
Oryzomys seuanezi		*												
Oryzomys subflavus		*	*							*				
Oryzomys talamancae					*	*								*

* species present P possible i introduced x extirpated	Argentina	Brazil	Bolivia	Chile	Colombia	Ecuador	French Guiana	Guyana	Paraguay	Peru	Suriname	Trinidad	Uruguay	Venezuela
Oryzomys tatei						*								
Oryzomys xantheolus						*				*				
Oryzomys yunganus		*	*		*	*	*	*		*	*			*
Oxymycterus akodontius	*													
Oxymycterus amazonicus		*												
Oxymycterus angularis		*												
Oxymycterus caparoae		*												
Oxymycterus dasytrichus		*												
Oxymycterus delator		*							*					
Oxymycterus hiska			*							*				
Oxymycterus hispidus		*												
Oxymycterus hucucha			*											
Oxymycterus inca			*							*				
Oxymycterus josei													*	
Oxymycterus nasutus		*											*	
Oxymycterus paramensis	*		*							*				
Oxymycterus quaestor	*	*							P					
Oxymycterus roberti		*												
Oxymycterus rufus	*													
Paralomys gerbillus										*				
Pearsonomys annectens				*										
Phaenomys ferrugineus		*												
Phyllotis amicus										*				
Phyllotis andium						*				*				
Phyllotis bonariensis	*													
Phyllotis caprinus	*		*											
Phyllotis darwini				*										
Phyllotis definitus										*				
Phyllotis haggardi						*								

	Argentina	Brazil	Bolivia	Chile	Colombia	Ecuador	French Guiana	Guyana	Paraguay	Peru	Suriname	Trinidad	Uruguay	Venezuela
* species present P possible i introduced x extirpated														
Phyllotis limatus				*						*				
Phyllotis magister				*						*				
Phyllotis osgoodi				*										
Phyllotis osilae	*		*							*				
Phyllotis wolffsohni			*											
Phyllotis xanthopygus	*			*						*				
Podoxymys roraimae		*						*						*
Pseudoryzomys simplex	*	*	*						*					
Punomys kofordi										*				
Punomys lemminus										*				
Reithrodon auritus	*			*										
Reithrodon tipicus	*	*											*	
Rhagomys longilingua										*				
Rhagomys rufescens		*												
Rhipidomys austrinus	*		*											
Rhipidomys caucensis					*									
Rhipidomys couesi					*							*		*
Rhipidomys emiliae		*												
Rhipidomys fulviventer					*									*
Rhipidomys gardneri		*								*				
Rhipidomys latimanus					*	*				*				
Rhipidomys leucodactylus		*	*		*	*	*	*			*			*
Rhipidomys macconnelli		*						*						*
Rhipidomys macrurus		*												
Rhipidomys masticalis		*												
Rhipidomys modicus										*				
Rhipidomys nitela		*				*	*				*			*
Rhipidomys ochrogaster										*				
Rhipidomys venezuelae					*							*		*

	Argentina	Brazil	Bolivia	Chile	Colombia	Ecuador	French Guiana	Guyana	Paraguay	Peru	Suriname	Trinidad	Uruguay	Venezuela
* species present P possible i introduced x extirpated														
Rhipidomys venustus														*
Rhipidomys wetzeli		*												*
Salinomys delicatus	*													
Scapteromys aquaticus	*								*					
Scapteromys tumidus		*											*	
Scolomys melanops						*				*				
Scolomys ucayalensis		*			*					*				
Sigmodon alstoni		*			*			*			*			*
Sigmodon hirsutus					*									*
Sigmodon inopinatus						*								
Sigmodon peruanus						*				*				
Sigmodontomys alfari					*	*								*
Sigmodontomys aphrastus						*								
Tapecomys primus			*											
Thalpomys cerradensis		*												
Thalpomys lasiotis		*												
Thaptomys nigrita	*	*							*					
Thomasomys apeco										*				
Thomasomys aureus			*		*	*				*				*
Thomasomys baeops						*								
Thomasomys bombycinus					*									
Thomasomys caudivarius						*								
Thomasomys cinereiventer					*									
Thomasomys cinereus										*				
Thomasomys cinnemeus						*								
Thomasomys daphne			*							*				
Thomasomys eleusis										*				
Thomasomys erro						*								
Thomasomys gracilis										*				

	Argentina	Brazil	Bolivia	Chile	Colombia	Ecuador	French Guiana	Guyana	Paraguay	Peru	Suriname	Trinidad	Uruguay	Venezuela
* species present P possible i introduced x extirpated														
Thomasomys hudsoni						*								
Thomasomys hylophilus					*									*
Thomasomys incanus										*				
Thomasomys ischyrus										*				
Thomasomys kalinowski										*				
Thomasomys ladewi			*											
Thomasomys laniger					*									*
Thomasomys macrotis										*				
Thomasomys monochromos					*									
Thomasomys niveipes					*									
Thomasomys notatus										*				
Thomasomys onkiro										*				
Thomasomys oreas			*							*				
Thomasomys paramorum						*								
Thomasomys popayanus					*									
Thomasomys praetor										*				
Thomasomys pyrrhonotus						*				*				
Thomasomys rhoadsi						*								
Thomasomys rosalinda										*				
Thomasomys silvestris						*								
Thomasomys taczanowskii			*							*				
Thomasomys ucucha						*								
Thomasomys vestitus														*
Thomasomys vulcani						*								
Weidomys pyrrhorhinos		*												
Wilfredomys oenax		*											*	
Zygodontomys brevicauda		*			*		*	*			*	*		*
Zygodontomys brunneus					*									

* species present P possible i introduced x extirpated	Argentina	Brazil	Bolivia	Chile	Colombia	Ecuador	French Guiana	Guyana	Paraguay	Peru	Suriname	Trinidad	Uruguay	Venezuela
Climbing rats														
Tylomys mirae					*	*								
Family Erethizontidae (Porcupines, etc.)														
Bristle-spined rat														
Chaetomys subspinosus		*												
Porcupines														
Coendou bicolor			*		*	*				*				
Coendou nycthemera		*												
Coendou prehensilis	*	*	*				*	*	*		*	*	*	*
Echinoprocta rufescens					*									
Sphiggurus ichillus						*								
Sphiggurus insidiosus		*												
Sphiggurus melanurus		*			*		*	*			*			P
Sphiggurus pruinosus					*									*
Sphiggurus roosmalenorum		*												
Sphiggurus spinosus	*	*							*				*	
Sphiggurus vestitus					*									
Sphiggurus villosus		*												
Family Chinchillidae (Chinchillas and vizcachas)														
Chinchilla chinchilla	*		*	*						*				
Chinchilla lanigera				*										
Lagidium peruanum				*						*				
Lagidium viscacia	*		*	*						*				
Lagidium wolffsohni	*			*										
Lagidium crassus										*				
Lagidium maximus	*		*						*					

* species present P possible i introduced x extirpated	Argentina	Brazil	Bolivia	Chile	Colombia	Ecuador	French Guiana	Guyana	Paraguay	Peru	Suriname	Trinidad	Uruguay	Venezuela
Family Dinomyidae (Pacarana)														
Dinomys branickii		*	*		*	*				*				*
Family Caviidae (Cavies, capybaras, etc.)														
Cavies														
Cavia aperea	*	*			*	*	*	*	*		*		*	*
Cavia fulgida		*												
Cavia intermedia		*												
Cavia magma		*											*	
Cavia tschudii	*		*	*						*				
Galea flavidens		*												
Galea musteloides	*		*	*						*				
Galea spixii		*	*											
Microcavia australis	*		P	*										
Microcavia niata			*	*										
Microcavia shiptoni	*													
Maras														
Dolichotis patagonum	*													
Dolichotis salinicola	*		*						*					
Capybaras and rock cavies														
Hydrochoerus hydrochaeris	*	*	*		*	*	*	*	*	*	*		*	*
Hydrochaeris isthmius					*									*
Kerodon acrobata		*												
Kerodon rupestris		*												
Family Dasyproctidae (Agoutis)														
Dasyprocta azarae	*	*							*					
Dasyprocta cristata							*	*			*			
Dasyprocta fuliginosa		*			*	*				*	*			*

	Argentina	Brazil	Bolivia	Chile	Colombia	Ecuador	French Guiana	Guyana	Paraguay	Peru	Suriname	Trinidad	Uruguay	Venezuela
* species present P possible i introduced x extirpated														
Dasyprocta guamara														*
Dasyprocta kalinowski										*				
Dasyprocta leporina		*					*	*			*			*
Dasyprocta prymnolopha		*												
Dasyprocta punctata	*	*	*		*	*				*				
Myoprocta acouchy		*					*	*			*			
Myoprocta pratti		*			*	*				*				*
Family Cuniculidae (Pacas)														
Cuniculus paca	*	*	*		*	*	*	*	*	*	*	*		*
Cuniculus taczanowskii					*	*				*				*
Family Ctenomyidae (Tuco-tucos)														
Ctenomys argentinus	*													
Ctenomys australis	*													
Ctenomys azarae	*													
Ctenomys bergi	*													
Ctenomys boliviensis	*		*						*					
Ctenomys bonettoi	*													
Ctenomys brasiliensis		*												
Ctenomys budini	*													
Ctenomys colburni	*													
Ctenomys coludo	*													
Ctenomys conoveri	*								*					
Ctenomys coyhaiquensis	P			*										
Ctenomys dorbigny	*													
Ctenomys dorsalis									*					
Ctenomys emilianus	*													
Ctenomys famosus	*													
Ctenomys flamarioni		*												

	Argentina	Brazil	Bolivia	Chile	Colombia	Ecuador	French Guiana	Guyana	Paraguay	Peru	Suriname	Trinidad	Uruguay	Venezuela
* species present P possible i introduced x extirpated														
Ctenomys fochi	*													
Ctenomys fodax	*													
Ctenomys frater	*		*											
Ctenomys fulvus	*			*										
Ctenomys goodfellowi			*											
Ctenomys haigi	*													
Ctenomys johannis	*													
Ctenomys juris	*													
Ctenomys knighti	*													
Ctenomys lami		*												
Ctenomys latro	*													
Ctenomys leucodon			*							*				
Ctenomys lewisi	P		*											
Ctenomys magellanicus	*			*										
Ctenomys maulinus	*			*										
Ctenomys mendocinus	*													
Ctenomys minutus	*	*											*	
Ctenomys nattereri		*												
Ctenomys occultus	*													
Ctenomys opimus	*		*	*						*				
Ctenomys osvaldoreigi	*													
Ctenomys pearsoni													*	
Ctenomys perrensis	*													
Ctenomys peruanus										*				
Ctenomys pilarensis									*					
Ctenomys pontifex	*													
Ctenomys porteousi	*													
Ctenomys pundti	*													
Ctenomys rionegrensis													*	
Ctenomys roigi	*													

* species present P possible i introduced x extirpated	Argentina	Brazil	Bolivia	Chile	Colombia	Ecuador	French Guiana	Guyana	Paraguay	Peru	Suriname	Trinidad	Uruguay	Venezuela
Ctenomys saltarius	*													
Ctenomys scagliai	*													
Ctenomys sericeus	*													
Ctenomys sociabilis	*													
Ctenomys steinbachi			*											
Ctenomys sylvanus	*													
Ctenomys talarum	*													
Ctenomys torquatus	*	*											*	
Ctenomys tuconax	*													
Ctenomys tucumanus	*													
Ctenomys tulduco	*													
Ctenomys validus	*													
Ctenomys viperinus	*													
Ctenomys yolandae	*													
Family Octodontidae (Degus)														
Aconaemys fuscus	*			*										
Aconaemys porteri	*			*										
Aconaemys sagei	*			P										
Octodon bridgesi				*										
Octodon degus				*										
Octodon lunatus				*										
Octodon pacificus				*										
Octodontomys gliroides	*		*	*										
Octomys mimax	*													
Pipanacoctomys aureus	*													
Salinoctomys loschalchalerosorum	*													
Spalacopus cyanus				*										
Tympanoctomys barrerae	*													

* species present P possible i introduced x extirpated	Argentina	Brazil	Bolivia	Chile	Colombia	Ecuador	French Guiana	Guyana	Paraguay	Peru	Suriname	Trinidad	Uruguay	Venezuela
Family Abrocomidae (Chinchilla rats)														
Abrocoma bennetti				*										
Abrocoma boliviensis			*											
Abrocoma budini	*													
Abrocoma cinerea	*		*	*						*				
Abrocoma famatina	*													
Abrocoma schistacea	*													
Abrocoma uspallata	*													
Abrocoma vaccarum	*													
Cuscomys ashaninka										*				
Cuscomys oblativa										*				
Family Echimyidae (Spiny rats, etc.)														
Bamboo rats														
Dactylomys boliviensis		*	*							*				
Dactylomys dactylinus		*	*		*	*				*				
Dactylomys peruanus										*				
Kannabateomys amblyonyx	*	*							*					
Olallamys albicauda					*									
Olallamys edax					*									*
Tree rats														
Callistomys pictus		*												
Diplomys caniceps					*	*								
Diplomys labilis					*	P								
Diplomys rufodorsalis					*									
Echimys chrysurus		*					*	*			*			
Echimys saturnus						*				*				
Echimys semivillosus					*									*
Isothrix bistriata		*			*	*				*				*

* species present P possible i introduced x extirpated	Argentina	Brazil	Bolivia	Chile	Colombia	Ecuador	French Guiana	Guyana	Paraguay	Peru	Suriname	Trinidad	Uruguay	Venezuela
Isothrix negrensis		*												
Isothrix pagurus		*												
Isothrix sinnamariensis							*							
Makalata didelphoides		*			*	*	*	*			*	*		*
Makalata grandis		*												
Makalata macrura		*				*				*				
Makalata obscura		*												
Makalata occasius						*				*				
Makalata rhipidura										*				
Phyllomys blainvillii		*												
Phyllomys brasiliensis		*												
Phyllomys dasythrix		*												
Phyllomys kerri		*												
Phyllomys lamarum		*												
Phyllomys lundi		*												
Phyllomys mantiqueirensis		*												
Phyllomys medius		*												
Phyllomys nigrispinus		*												
Phyllomys pattoni		*												
Phyllomys thomasi		*												
Phyllomys unicolor		*												
Spiny rats														
Carterodon sulcidens		*												
Clyomys bishopi		*												
Clyomys laticeps		*												
Euryzygomatomys spinosus	*	*							*					
Hoplomys gymnurus						*								
Lonchothrix emiliae		*												
Mesomys hispidus		*			*	*	*	*		*	*			*
Mesomys leniceps										*				

	Argentina	Brazil	Bolivia	Chile	Colombia	Ecuador	French Guiana	Guyana	Paraguay	Peru	Suriname	Trinidad	Uruguay	Venezuela
* species present P possible i introduced x extirpated														
Mesomys occultus		*												
Mesomys stimulax		*												
Proechimys brevicauda		*	*		*	*				*				
Proechimys canicollis					*									*
Proechimys chrysaeolus					*									
Proechimys cuvieri		*					*	*		*	*			
Proechimys decumanus						*				*				
Proechimys echinothrix		*			P									
Proechimys gardneri		*	*											
Proechimys goeldii		*												
Proechimys guairae														*
Proechimys guyannensis		*					*	*			*			*
Proechimys hoplomyoides		*						*						*
Proechimys kulinae		*								*				
Proechimys longicaudatus		*	*						*					
Proechimys magdalenae					*									
Proechimys mincae					*									
Proechimys oconnelli					*									
Proechimys pattoni		*								*				
Proechimys poliopus					*									*
Proechimys quadruplicatus		*			*	*				*				*
Proechimys roberti		*												
Proechimys semispinosus					*	*								
Proechimys simonsi		*	*		*	*				*				
Proechimys steerei		*	*							*				
Proechimys trinitatus												*		
Proechimys urichi														*
Thrichomys apereoides		*												
Thrichomys inermis		*												
Thrichomys pachyurus		*							*					

* species present P possible i introduced x extirpated	Argentina	Brazil	Bolivia	Chile	Colombia	Ecuador	French Guiana	Guyana	Paraguay	Peru	Suriname	Trinidad	Uruguay	Venezuela
Trinomys albispinus		*												
Trinomys dimidiatus		*												
Trinomys elasi		*												
Trinomys gratiosus		*												
Trinomys iheringi		*												
Trinomys mirapitanga		*												
Trinomys moojeni		*												
Trinomys myosuros		*												
Trinomys paratus		*												
Trinomys setosus		*												
Trinomys yonenagae		*												
Family Myocastoridae (Coypu)														
Myocastor coypus	*	*	*	*					*				*	
Order Lagomorpha (Hares and rabbits)														
Family Leporidae														
Lepus europeus	i			i										
Oryoctolagus cuniculus	i			i										
Sylvilagus brasiliensis	*	*	*		*	*			*	*				*
Sylvilagus floridanus					*									*
Order Soricomorpha (Shrews and moles)														
Family Soricidae (Shrews)														
Cryptotis brachyonyx					*									
Cryptotis colombiana					*									
Cryptotis equatoris						*								
Cryptotis medellinius					*									
Cryptotis meridensis														*
Cryptotis merus					*									
Cryptotis montivagus						*								

* species present P possible i introduced x extirpated	Argentina	Brazil	Bolivia	Chile	Colombia	Ecuador	French Guiana	Guyana	Paraguay	Peru	Suriname	Trinidad	Uruguay	Venezuela
Cryptotis peruviensis										*				
Cryptotis squamipes					*	*								
Cryptotis tamensis														*
Cryptotis thomasi					*	*				*				
Order Chiroptera (Bats)														
Family Emballonuridae (Sac-winged bats)														
Balantiopteryx infusca					*	*								
Balantiopteryx plicata					*									
Centronycteris centralis					*	*				*				
Centronycteris maximiliani		*					*	*		*	*			*
Cormura brevirostris		*			*	*				*				
Cyttarops alecto		*					*	*						
Diclidurus albus		*			*	*				*	*	*		*
Diclidurus ingens		*			*			*						*
Diclidurus isabellus		*						*						*
Diclidurus scutatus		*					*	*		*	*			*
Peropteryx kappleri		*	*		*	*	*	*		*	*			*
Peropteryx leucoptera		*			*	*	*	*		*	*			*
Peropteryx macrotis		*			*		*	*	*		*			*
Peropteryx trinitatus							*					*		*
Rhynchonycteris naso		*			*	*	*	*		*	*	*		*
Saccopteryx antioquensis					*									
Saccopteryx bilineata		*	*		*	*	*	*		*	*	*		*
Saccopteryx canescens		*	*		*	*	*	*		*	*			*
Saccopteryx gymnura		*					*	*						P
Saccopteryx leptura		*	*		*	*	*	*		*	*	*		*
Family Phyllostomidae (Leaf-nosed bats)														

* species present P possible i introduced x extirpated	Argentina	Brazil	Bolivia	Chile	Colombia	Ecuador	French Guiana	Guyana	Paraguay	Peru	Suriname	Trinidad	Uruguay	Venezuela
Vampire bats														
Desmodus rotundus	*	*	*	*	*	*	*	*	*	*	*	*	*	*
Diaemus youngi	*	*	*		*	*	*	*	*	*	*	*		*
Diphylla ecaudata		*	*		*	*	*	*		*	*			*
Tailless, long-tailed and long-tongued bats														
Anoura caudifer	*	*	*		*	*	*	*		*	*			*
Anoura cultrata			*		*	*				*				*
Anoura fistulata					P	*				P				
Anoura geoffroyi	*	*	*		*	*	*	*		*	*	*		*
Anoura latidans					*			*		*				*
Anoura luismanueli														*
Choeroniscus godmani					*		*	*			*	*		*
Choeroniscus minor		*	*		*	*	*	*		*	*	*		*
Choeroniscus periosus					*	*								*
Glossophaga commissarisi		*			*	*				*				
Glossophaga longirostris		*			*			*				*		*
Glossophaga soricina	*	*	*		*	*	*	*	*	*	*	*		*
Leptonycteris curasoae					*							*		*
Lichonycteris obscura		*	*		*			*			*	*		*
Scleronycteris ega		*												*
Nectar bats														
Lionycteris spurrelli		*	*		*		*	*		*	*			*
Lonchophylla bokkermanni		*												
Lonchophylla dekeyseri		*												
Lonchophylla handleyi					*	*				*				
Lonchophylla hesperia						*				*				
Lonchophylla mordax		*	*		*	*				*				
Lonchophylla robusta					*	*				*				*
Lonchophylla thomasi		*			*	*	*	*		*	*			*

* species present P possible i introduced x extirpated	Argentina	Brazil	Bolivia	Chile	Colombia	Ecuador	French Guiana	Guyana	Paraguay	Peru	Suriname	Trinidad	Uruguay	Venezuela
Platalina genovensium										*				
Xeronycteris vieirai		*												
Sword-nosed bats														
Chrotopterus auritus	*	*	*		*	*	*	*	*	*	*			*
Glyphonycteris behnii		*												
Glyphonycteris daviesi		*	*		*	*				*				
Glyphonycteris sylvestris		*			*					*	*	*		*
Lampronycteris brachyotis		*			*		*	*		*	*			*
Lonchorhina aurita		*	*		*	*	*	*		*	*			*
Lonchorhina fernandezi														*
Lonchorhina inusitata		*					*	*			*			*
Lonchorhina marinkellei					*									
Lonchorhina orinocensis					*									*
Lophostoma brasiliense		*	*		*	*	*	*		*	*	*		*
Lophostoma carrekeri		*	*		*		*	*		*	*			*
Lophostoma schulzi		*					*	*			*			
Lophostoma silvicolum		*	*		*	*	*	*	P	*	*	*		*
Macrophyllum macrophyllum	*	*	*		*	*	*	*	*	*	*			*
Micronycteris brosseti		*					*	*		*				
Micronycteris hirsuta		*			*	*	*	*		*	*	*		*
Micronycteris homezi		*					*	*						*
Micronycteris matses										*				
Micronycteris megalotis		*			*	*	*	*		*	*	*		*
Micronycteris microtis		*	*		*		*	*		*	*			*
Micronycteris minuta		*	*		*	*	*	*		*	*	*		*
Micronycteris sanborni		*	*											
Micronycteris schmidtorum		*			*					*				*
Mimon bennettii		*			*		*	*			*			*
Mimon cozumelae					*									
Mimon crenulatum		*	*		*	*	*	*		*	*	*		*

* species present P possible i introduced x extirpated	Argentina	Brazil	Bolivia	Chile	Colombia	Ecuador	French Guiana	Guyana	Paraguay	Peru	Suriname	Trinidad	Uruguay	Venezuela
Mimon koepckeae										*				
Neonycteris pusilla		*			*									
Phylloderma stenops		*	*		*	*	*	*		*	*	*		*
Phyllostomus discolor		*	*		*	*	*	*	*	*	*	*		*
Phyllostomus elongatus		*	*		*	*	*	*		*	*	*		*
Phyllostomus hastatus	*	*	*		*	*	*	*	*	*	*	*		*
Phyllostomus latifolus		*			*		*	*			*			*
Tonatia bidens	*	*	*						*					
Tonatia saurophila		*			*		*	*		*	*	*		*
Trachops cirrhosus		*	*		*	*	*	*		*	*	*		*
Trinycteris nicefori		*	*		*	*	*	*		*	*	*		*
Vampyrum spectrum		*	*		*	*	*	*		*	*	*		*
Short-tailed bats														
Carollia brevicauda		*	*		*	*	*	*		*	*	*		*
Carollia castanea		*	*		*	*	*	*		*	*			*
Carollia colombiana					*									
Carollia perspicillata	*	*	*		*	*	*	*	*	*	*	*		*
Rhinophylla alethina					*	*								
Rhinophylla fischerae		*			*	*				*				*
Rhinophylla pumilio		*	*		*	*	*	*		*	*			*
Yellow-shouldered bats														
Sturnira aratathomasi					*	*				*				*
Sturnira bidens		P			*	*				*				*
Sturnira bogotensis					*	*				*				
Sturnira erythromos	*		*		*	*				*				*
Sturnira lilium	*	*	*		*	*	*	*	*	*	*	*	*	*
Sturnira ludovici					*	*		*		*		*		*
Sturnira luisi					*	*				*				
Sturnira magna		*	*		*	*				*				

* species present P possible i introduced x extirpated	Argentina	Brazil	Bolivia	Chile	Colombia	Ecuador	French Guiana	Guyana	Paraguay	Peru	Suriname	Trinidad	Uruguay	Venezuela
Sturnira mistratensis					*									
Sturnira nana										*				
Sturnira oporaphilum	*		*			*				*				
Sturnira tildae		*	*		*	*	*	*		*	*	*		*
Fruit-eating bats														
Ametrida centurio		*					*	*			*	*		*
Artibeus amplus					*			*						*
Artibeus anderseni		*	*			*				*				
Artibeus cinereus		*					*	*		*	*			*
Artibeus concolor					*	*	*	*		*	*			*
Artibeus fimbriatus		*							*					
Artibeus fraterculus						*				*				
Artibeus glaucus		*	*		*	*	*	*		*	*	*		*
Artibeus gnomus		*	*		*	*	*	*		*	*			*
Artibeus jamaicensis	*	*	*		*	*	*	*	*	*	*	*		*
Artibeus lituratus	*	*	*		*	*	*	*	*	*	*	*		*
Artibeus obscurus		*	*		*	*	*	*		*	*			*
Artibeus phaeotis						*		*		*				
Artibeus watsoni					*									
Centurio senex					*							*		*
Chiroderma doriae		*							*					
Chiroderma salvini			*		*	*				*				*
Chiroderma trinitatum		*	*		*	*	*	*		*	*	*		*
Chiroderma villosum		*	*		*	*	*	*		*	*	*		*
Enchisthenes hartii			*		*	*		*				*		*
Mesophylla macconnelli		*	*		*	*	*	*		*	*	*		*
Platyrrhinus aurarius								*			*			*
Platyrrhinus brachycephalus		*	*		*	*	*	*		*	*			*
Platyrrhinus chocoensis					*	*								

	Argentina	Brazil	Bolivia	Chile	Colombia	Ecuador	French Guiana	Guyana	Paraguay	Peru	Suriname	Trinidad	Uruguay	Venezuela
* species present P possible i introduced x extirpated														
Platyrrhinus dorsalis			*		*	*				*				
Platyrrhinus helleri		*	*		*	*	*	*		*	*	*		*
Platyrrhinus infuscus		*	*		*	*				*				
Platyrrhinus lineatus	*	*	*		*	*	*			*	*		*	
Platyrrhinus recifinus		*												
Platyrrhinus umbratus					*									*
Platyrrhinus vittatus			*		*	*				*				*
Pygoderma bilabiatum	*	*	*						*					
Sphaeronycteris toxophyllum		*	*		*	*		*		*				*
Uroderma bilobatum		*	*		*	*	*	*		*	*	*		*
Uroderma magnirostrum		*	*		*	*				*				*
Vampressa bidens		*	*		*	*	*	*		*	*			*
Vampressa brocki		*			*		*	*		*	*			
Vampressa melissa					*	P				*				
Vampressa nymphaea					*	*								
Vampressa pusilla		*	*		*	*	*	*		*	*	*		*
Vampressa thyone			*				*	*		*				
Vampyrodes caraccioli		*	*		*	*	*	*		*	*	*		*
Family Mormoopidae (Leaf-chinned bats)														
Mormoops megalophylla					P	*				*		*		*
Pteronotus davyi										*				*
Pteronotus gymnonotus		*	*				*	*		*				
Pteronotus parnellii		*	*				*	*		*	*	*		*
Pteronotus personatus		*	*		*		*	*		*	*	*		*
Family Noctilionidae (Bulldog or fishing bats)														
Noctilio albiventris	*	*	*		*	*	*	*		*	*	*		*
Noctilio leporinus	*	*	*		*	*	*	*	*	*	*	*		*

	Argentina	Brazil	Bolivia	Chile	Colombia	Ecuador	French Guiana	Guyana	Paraguay	Peru	Suriname	Trinidad	Uruguay	Venezuela
* species present P possible i introduced x extirpated														
Family Furipteridae (Smoky bats)														
Amorphochilus schnablii				*		*				*				
Furipterus horrens		*			*		*	*		*	*	*		*
Family Thyropteridae (Disc-winged bats)														
Thyroptera discifera		*	*		*	*	*	*		*	*			*
Thyroptera lavali		*				*				*				*
Thyroptera tricolor		*	*		*	*	*	*		*	*	*		*
Family Natalidae (Funnel-eared bats)														
Natalus stramineus		*	*		*		*	*			*			*
Natalus tumidirostris					*							*		*
Family Molossidae (Free-tailed bats)														
Blunt-eared bat														
Tomopeas ravus										*				
Free-tailed bats														
Cynomops abrasus	*	*	*		*		*	*	*	*	*			*
Cynomops greenhalli		*				*	*	*		*	*	*		*
Cynomops paranus	*	*			*	*	*	*		*	*			*
Cynomops planirostris		*	*		*	P	*	P	*	*	*			*
Eumops auripendulus	*	*	*		*	*	*	*		*	*	*		*
Eumops bonariensis	*	*	*		*	P	*	*	*	*	*		*	*
Eumops dabbnei	*	*			*				*					*
Eumops glaucinus	*	*	*		*	*	*	*	*	*	*			*
Eumops hansae		*	*		*		*	*		*	*			*
Eumops maurus						*		*			*			*
Eumops perotis	*	*	*		*	*			*	*				*
Eumops trumbulli		*	*		*		*	*		*	*			*
Molossops aequatorianus						*								
Molossops mattogrossensis		*						*						*

* species present P possible i introduced x extirpated	Argentina	Brazil	Bolivia	Chile	Colombia	Ecuador	French Guiana	Guyana	Paraguay	Peru	Suriname	Trinidad	Uruguay	Venezuela
Molossops neglectus	*	*			*			*		*	*			*
Molossops temminckii	*	*	*		*	*		*	*	*			*	*
Molossus barnesi							*							
Molossus coibensis		*			*	*		*		*				*
Molossus currentium	*	*			*	*			*					*
Molossus molossus	*	*	*		*	*	*	*	*	*	*	*	*	*
Molossus pretiosus		*			*			*						*
Molossus rufus	*	*			*	*	*	*		*	*	*		*
Molossus sinaloae					*		*	*			*	*		
Mormopterus kalinowski				*						*				
Mormopterus phrudus										*				
Nyctinomops aurispinosus		*	*							*				
Nyctinomops laticaudatus	*	*	*		*		*	*	*	*	*	*		*
Nyctinomops macrotis	*	*	*		*			*		*	*		*	*
Promops centralis	*	*	*		*		*	*	*	*	*			*
Promops nasutus	*	*	*			*		*	*	*	*	*		*
Tadarida brasiliensis	*	*	*	*	*	*	*	*	*	*	*	*	*	*
Family Vespertilionidae (Vesper bats)														
Serotine bats														
Eptesicus andinus		*	P		*	*				*				*
Eptesicus brasiliensis	*	*	*		*	*	*	*	*	*	*	*	*	*
Eptesicus chiriquinus					*	*	*	*		*	*			*
Eptesicus diminutus	*	*							*				*	*
Eptesicus furinalis	*	*			*		*	*			*	*		*
Eptesicus fuscus		*			*									*
Eptesicus innoxius						*				*				
Hairy-tailed bats														
Lasiurus atratus							*	*			*			*
Lasiurus blossevillii	*	*	*	*	*	*	*	*	*	*	*	*	*	*

* species present P possible i introduced x extirpated	Argentina	Brazil	Bolivia	Chile	Colombia	Ecuador	French Guiana	Guyana	Paraguay	Peru	Suriname	Trinidad	Uruguay	Venezuela
Lasiurus cinereus	*		*	*	*								*	*
Lasiurus ebanus		*												
Lasiurus ega	*	*	*		*	*	*	*	*	*	*	*	*	*
Lasiurus egregius		*					*							
Lasiurus varius	*			*										
Rhogeesa io		*	*		*	*		*				*		*
Rhogeesa minutilla					*									*
Big-eared bats														
Histiotus alienus		*											*	
Histiotus humboldti					*									*
Histiotus laephotis	*		*							*				
Histiotus macrotus	*			*										
Histiotus magellicanus	*			*										
Histiotus montanus	*	P	*	*	*	*				*			*	*
Histiotus velatus	*	*	*						*					
Little brown bats														
Myotis aelleni	*													
Myotis albescens	*	*	*		*	*		*	*	*	*		*	*
Myotis atacamensis				*						*				
Myotis chiloensis	*			*										
Myotis keaysi	*	*	*		*	*				*		*		*
Myotis levis	*	*	*										*	
Myotis nesopolus														*
Myotis nigricans	*	*	*		*	*	*	*	*	*	*	*	*	*
Myotis oxyotus		*			*	*								*
Myotis riparius	*	*	*		*		*	*	*		*	*	*	*
Myotis ruber	*	*							*					
Myotis simus	*	*	*		*	*				*				

* species present P possible i introduced x extirpated	Argentina	Brazil	Bolivia	Chile	Colombia	Ecuador	French Guiana	Guyana	Paraguay	Peru	Suriname	Trinidad	Uruguay	Venezuela
Order Carnivora (Carnivores)														
Family Felidae (Cats)														
Leopardus braccatus		*							*				*	
Leopardus colocolo				*										
Leopardus geoffroyi	*	*	*	*					*				*	
Leopardus guigna	*			*										
Leopardus jacobita	*		*	*						*				
Leopardus pajeros	*		*	*		*				*				
Leopardus pardalis	*	*	*		*	*	*	*	*	*	*	*	*	*
Leopardus tigrinus	*	*	*		*	*	*	*	*	*	*			*
Leopardus weidii	*	*	*		*	*	*	*	*	*	*		*	*
Panthera onca	*	*	*		*	*	*	*	*	*	*	*	*	*
Puma concolor	*	*	*	*	*	*	*	*	*	*	*	*	*	*
Puma yagouaroundi	*	*	*		*	*	*	*	*	*	*	*	*	*
Family Herpestidae (Mongoose)														
Herpestes javanicus											i			
Family Canidae (Dogs)														
Atelocynus microtis		*			*	*				*				
Cerdocyon thous	*	*	*		*		*	*	*	*	*		*	*
Chrysocyon brachyurus	*	*	*						*				*	
Lycalopex culpaeus	*		*	*	*	*				*				
Lycalopex fulvipes				*										
Lycalopex griseus	*			*										
Lycalopex gymnocercus	*	*	*						*				*	
Lycalopex sechurae						*				*				
Lycalopex vetulus		*												
Speothos venaticus		*	*		*	*	*	*	*	*	*			*
Urocyon cinereoargenteus					*									*

	Argentina	Brazil	Bolivia	Chile	Colombia	Ecuador	French Guiana	Guyana	Paraguay	Peru	Suriname	Trinidad	Uruguay	Venezuela
* species present P possible i introduced x extirpated														
Family Ursidae (Spectacled bear)														
Tremarctos ornatus			*		*	*				*				*
Family Otaridae (Eared seals)														
Arctocephalus australis	*	*		*						*			*	
Arctocephalus philippii				*										
Otaria flavescens	*	*		*						*			*	
Family Phocidae (Earless seals)														
Hydrurga leptonyx				*										
Leptonychotes weddellii	*			*										
Lobodon carcinophaga	*	*											*	
Mirounga leonina	*	*		*									*	
Family Mustelidae (Weasels and otters)														
Lontra felina	*			*						*				
Lontra longicaudis	*	*			*	*	*	*		*	*			*
Lontra provocax	*			*										
Pteronura brasiliensis	*	*	*		*	*	*	*	*	*	*		*	*
Weasels														
Eira barbara	*	*	*		*	*	*	*	*	*	*	*	*	*
Galictis cuja	*	*	*	*					*	*				
Galictis vittata	*		*		*					*				*
Lyncodon patagonicus	*			*										
Mustela africana		*				*				*				
Mustela felipei					*	*								
Mustela frenata					*					*				*
Neovison vison	i			i										

* species present P possible i introduced x extirpated	Argentina	Brazil	Bolivia	Chile	Colombia	Ecuador	French Guiana	Guyana	Paraguay	Peru	Suriname	Trinidad	Uruguay	Venezuela
Family Mephitidae (Skunks)														
Conepatus chinga	*	*	*	*						*			*	
Conepatus humboldtii	*								*					
Conepatus semistriatus		*			*	*				*				*
Family Procyonidae (Raccoons, etc.)														
Olingos and kinkajou														
Bassaricyon alleni			*			*				*				
Bassaricyon beddardi		*						*						*
Bassaricyon gabbii					*	*								*
Petos flavus		*	*		*	*	*	*		*	*			*
Coatis and raccoon														
Nasua narica					*					*				
Nasua nasua	*	*	*		*			*	*	*	*		*	*
Nasuella olivacea					*	*								*
Procyon cancrivorus	*	*	*		*	*			*		*	*		*
Order Perrisodactyla (Odd-toed ungulates)														
Family Tapiridae (Tapirs)														
Tapirus bairdii					*	*								
Tapirus pinchaque					*	*				P				P
Tapirus terrestris	*	*	*		*	*	*	*	*	*	*			*
Order Arteriodactyla (Even-toed ungulates)														
Family Suidae (European or Wild Boar)														
Sus scrofa	i			i										
Family Tayassuidae (Peccaries)														
Catagonus wagneri	*		*						*					

* species present P possible i introduced x extirpated	Argentina	Brazil	Bolivia	Chile	Colombia	Ecuador	French Guiana	Guyana	Paraguay	Peru	Suriname	Trinidad	Uruguay	Venezuela
Pecari tajacu	*	*	*		*	*	*	*	*	*	*			*
Tayassu pecari	*	*	*		*	*	*	*	*	*	*			*
Family Camelidae (Llamas and vicuña)														
Llama glama (Llama)	*		*	*	i	*				*				i
Llama glama (Guanaco)	*			*										
Llama glama (Alpaca)	*		*	*		*				*				
Vicugna vicugna	*		*	*						*				
Family Cervidae (Deer)														
Blastocerus dichotomus	*	*	*						*	*			*	
Cervus elaphas	i			i										
Hippocamelus antisensis	*		*			*				*				
Hippocamelus bisulcus	*			*										
Mazama americana	*	*	*		*		*	*		*	*	*		*
Mazama bororo		*												
Mazama bricenii														*
Mazama chunyi			*							*				
Mazama gouazoubira	*	*	*		*	*	*	*	*	*	*		*	*
Mazama nana	*	*							*					
Mazama rufina					*	*								
Mazama temama					*									
Odocoileus virginianus		*	*		*		*	*		*	*			*
Ozotoceros bezoarticus	*	*	*						*				*	
Pudu mephistophiles					*	*				*				
Pudu puda	*			*										
Family Bovidae (Blackbuck)														
Antilope cervicapra	i													

	Argentina	Brazil	Bolivia	Chile	Colombia	Ecuador	French Guiana	Guyana	Paraguay	Peru	Suriname	Trinidad	Uruguay	Venezuela
* species present P possible i introduced x extirpated														
Order Cetacea (Whales and dolphins, inland species only)														
Family Delphinidae (Dolphins)														
Sotalia fluviatilis		*			*	*	*	*		*	*			*
Family Iniidae (River dolphins)														
Inia geoffrensis		*	*		*	*				*				*

REFERENCES

Albuja-Viteri, Luís. 1999. Murciélagos del Ecuador. Segunda Edición. Quito, Ecuador: Escuela Politécnica Nacional. 288 pp.

Baker, Robert J., Knox Jones Jr., and Dilford C. Carter. 1977. Biology of the bats of the New World Family Phyllostomatidae. Parts I, II, and III. Special Publications. Lubbock, TX: The Museum, Texas Tech University. 218 pp., 364 pp., 441 pp.

Barquez, Rubén M., Norberto P. Giannini, and Michael A. Mares. 1993. Bats of Argentina. Norman, OK: Oklahoma Museum of Natural History. 119 pp.

Barquez, Rubén M., Michael A. Mares, and Ricardo A. Ojeda. 1991. Mammals of Tucuman. Norman, OK: Oklahoma Museum of Natural History. 282 pp.

Bonino, Never. 2005. Guía de Mamíferos de la Patagonia Argentina. Bariloche, Argentina: Instituto Nacional de Tecnología Agropecuaria (INTA), 112 pp.

Cabrera, Angel. 1957. Catálogo de mamíferos de América del Sur. Rev. Mus. Argent. Cienc. Nat. Bernardino Rivadavia Inst. Nac. Invest. Cienc. Nat. (Argent.). Tomo 4, no. 1. Buenos Aires Imprenta e Casa Editora Con: 634, Pere, 684.

Cabrera, Angel, and José Yepes. 1960. Mamiferos Sud Americanos. 2 vols. Buenos Aires: Ediar, S.A. 347 pp.

Carlton, M. D., and G. G. Musser. 1989. Systematic studies of oryzomine rodents (Muridae, Sigmodontinae) a synopsis of microoryzomys. Bull. Am. Mus. Nat. Hist. 191:1–83.

Davis, William B. 1969. A review of the small fruit bats (genus Artibeus) of Middle America. Southwestern Nat. 14:15–29.

Davis, William B. 1984. A review of the large fruit-eating bats of the Artibeus lituratus complex (Chiroptera: Phyllostomatidae) in Middle America. Occasional Papers. Lubbock, TX: The Museum, Texas Tech University. 93:1–16.

Eisenberg, John F. 1989. The northern neotropics, Vol. 1 of Mammals of the neotropics. Chicago: University of Chicago Press. 449 pp.

Eisenberg, John F., and Kent H. Redford. 1999. The central neotropics, Vol. 3 of Mammals of the neotropics. Chicago: University of Chicago Press. 609 pp.

Emmons, Louise H. 1990. Neotropical rainforest mammals. Chicago: University of Chicago Press. 281 pp.

Fernandez-Badillo, Alberto, Ricardo Guerrero, Rexford Lord, José Ochoa, and Gregorio Ulloa. 1988. Mamíferos en Venezuela, Lista y Claves para su Identificación. Caracas: Facultad de Agronomía, Universidad Central de Venezuela. 185 pp.

Glydenstolpe, N. 1932. A manual of neotropical Sigmodont rodents. Kungl. Svensk. Vétenskapsakad. Handl. 11:1–164.

Goodwin, G. G., and Arthur M. Greenhall. 1961. A review of the bats of Trinidad and Tobago. Bull. Mus. Nat. Hist. 122 (3): 187–302.

Greenhall, Arthur M., Rexford D. Lord, and Elio Massoia. 1983. Key to the bats of Argentina. Buenos Aires: Pan American Zoonosis Center, Pan American Health Organization, Ramos Mejia. 103 pp.

Handley, Charles O., Jr. 1976. Mammals of the Smithsonian Venezuelan Project. Brigham Young Univ. Sci. Bull. Biol. Ser. 20 (5): 1–90.

Harris, Graham 1998. A guide to the birds and mammals of coastal Patagonia. Princeton, NJ: Princeton University Press. 231 pp.

Hershkovitz, Philip. 1962. Evolution of Neotropical cricetine rodents (Muridae). Vol. 46 of Fieldiana zoology. Chicago: Field Museum of Natural History. 524 pp.

Hershkovitz, Philip. 1977. Living New World monkeys (Platyrrhini). Vol. 1. Chicago: University of Chicago Press. 1,117 pp.

Hornacki, James H., K. Kinman, and J. Koeppl, eds. 1982. Mammal Species of the world. Lawrence, KS: Allen Press, Inc. and Association of Systematics Collections. 694 pp.

Husson, A.M. 1978. The mammals of Suriname. Leiden, The Netherlands: E. J. Brill. 569 pp.

Linares, Omar, J. 1987. Murciélagos de Venezuela. Caracas: Cuadernos Lagoven. 120 pp.

Linares, Omar J. 1998. Mamíferos de Venezuela. Caracas: Sociedad Conservacioista Audobon, de Venezuela, Universidad Simón Bolivar. 691 pp.

Lord, Rexford D. 1999. Wild mammals of Venezuela. Caracas: Armitano Editores, C.A. 347 pp.

Mann-Fischer, Guillermo. 1978. Los Pequeños mamíferos de Chile. Gayana, no. 40. Santiago: Universidad de Concepción. 342 pp.

Mares, Michael A., and Hugh H. Genoways, eds. 1982. Mammalian biology in South America. Vol. 6. Pymatuning. Lineville, PA: Laboratory of Ecology, University of Pittsburgh. 539 pp.

Mares, Michael A., Ricardo A. Ojeda, and Rubén M. Barquez. 1989. Guide to the mammals of Salta Province, Argentina. Norman, OK: University of Oklahoma Press. 303 pp.

Montgomery, Gene G., ed. 1985. The evolution and ecology of armadillos, sloths and vermilinguas. Washington, DC: Smithsonian Institution Press. 451 pp.

Musser, Guy G., Michael D. Carleton, Eric M. Brothers, and Alfred L. Gardner. 1998. Systematic studies of Oryzomine rodents (Muridae, Sigmodontinae):

Diagnoses and distributions of species formerly assigned to Oryzomys "capito." Bull. Am. Mus. Nat. Hist. Number 236, 376 pp.

Nowak, Ronald M. 1999. Walker's mammals of the world, 6th ed. 2 vols. Baltimore: Johns Hopkins University Press. 1,935 pp.

Olrog, C. C., and M. M. Lucero. 1981. Guia de los Mamíferos Argentinos. Tucuman, Argentina: Fundación Miguel Lillo. 151 pp.

Pérez-Hernandez, Roger, Pascual Soriano, and Daniel Lew. 1994. Marsupiales de Venezuela. Caracas: Editorial Arte, S.A., Cuadernos Lagoven. 76 pp.

Redford, Kent H., and John F. Eisenberg. 1989. Advances in neotropical mammalogy. Gainesville, FL: Sandhill Crane Press. 613 pp.

Redford, Kent H., and John F. Eisenberg. 1992. The southern cone. Vol. 2 of Mammals of the neotropics. Chicago: University of Chicago Press. 430 pp.

Reid, Fiona A. 1997. Mammals of Central America and Southeast Mexico. New York: Oxford University Press. 334 pp.

Rylands, Anthony B. 1993. Marmosets and tamarins. Systematics, behavior and ecology. New York: Oxford University Press. 396 pp.

Tirira S., Diego. 1999. Mamíferos del Ecuador. Publicación Especial 2. Quito, Ecuador: Museo de Zoología, Centro de Biodiversidad y Ambiente, Pontifica Universidad Católica del Ecuador.

Wilson, Don E., and F. Russel Cole. 2000. Common names of mammals of the world. Washington, DC: Smithsonian Institution Press. 1,206 pp.

Wilson, Don E., and DeeAnn M. Reeder, eds. 2005. Mammal species of the world, 3rd ed. 2 vols. Baltimore: Johns Hopkins University Press. 2,142 pp.

INDEX